HIGH TIDE

mark lynas

HARPER PERENNIAL

Harper Perennial
nt of HarperCollins*Publishers*
85 Fulham Palace Road
Hammersmith
London W6 8JB

www.harperperennial.co.uk

This edition published by Harper Perennial 2005
1

First published by Flamingo in 2004

ISBN 0 00 713940 3

Typeset in Postscript Baskerville Classico by
Rowland Phototypesetting Ltd
Bury St Edmunds, Suffolk

Printed and bound in Great Britain by
Clays Ltd, St Ives plc

Mark Lynas was born in Fiji in 1973 and grew up in Peru, Spain and the UK. He studied history and politics at Edinburgh University, then edited the OneWorld.net website until 2000. A specialist in climate change, Mark is now a journalist, campaigner and commentator. He lives in Oxford. Visit his website: www.marklynas.org

For automatic updates on Mark Lynas visit HarperPerennial.co.uk and register for AuthorTracker

'Lynas's book is a clarion call to action ... clear, lucid and informative. Global warming is already happening, with catastrophic results for humanity. Lynas's book amply demonstrates this, to the satisfaction of even the most apocalyptic among us.' WILL SELF, *New Statesman*

'Lynas tells us to keep repeating the climate change message. Read this book and that is exactly what you will do.'
MICHAEL MEACHER, *Guardian*

'Like some Green Orwell, Lynas spent three years looking for evidence of how climate change is starting to impact on everyday lives. He found plenty.' *Independent*

'A harrowing prospect, brilliantly set out ... Lynas paints a fearsome picture of the near future.' J. G. BALLARD

'Even if you have never read another green book in your life this one is highly recommended.' *New Humanist*

For my family

Contents

Acknowledgements

Although most of the information in this book is based on the accounts of ordinary people and on my own experiences, its claim to rigour and truth is ultimately founded on the work of hundreds of climatologists, meteorologists, atmospheric physicists and other scientific experts, very few of whom ever receive one tenth of the recognition that they deserve. In an age where science has often been perverted in the interests of the moneyed and the powerful, atmospheric science has retained a 'blue skies' element in all senses of the phrase, and is much the richer for it. A skim through the endnotes will give the reader the names of just a few of the experts whom I specifically cite (and without whose original work my analysis would have been impossible), but many more remain unacknowledged for their labours in contributing to the current solid consensus about the reality of climate change.

I have benefited immensely from an informal scientific 'peer review' process of early drafts of this manuscript, although if any errors either of fact or interpretation remain then the responsibility for them rests squarely on my shoulders alone. I am particularly indebted to Sir John Houghton, one of the most eminent climate scientists in the UK and author of *Global Warming: The Complete Briefing* for his comments. I am also deeply grateful that Dr Rajendra Pachauri managed to take time out of his busy schedule as Chairman of the Intergovernmental Panel on Climate Change to send me extensive and very valuable feedback.

Specific chapters were also reviewed by experts in the relevant fields, many of whom you'll meet in the chapters themselves. Again, I am deeply in their debt. Specifically, they are: Dr Tim Osborn, Senior Research Associate at the University of East Anglia's Climatic Research Unit on Chapter 1; Professor Gunter Weller, Director of the Co-operative Institute of Arctic Research at the University of Alaska Fairbanks on Chapter 2; Patrick Nunn, Professor of Oceanic Geoscience and Head of the Department of Geography at the University of the South Pacific on Chapter 3; Professor Zong-ci Zhao at the China Meteorological Administration in Beijing on Chapter 4; Tom Knutson at the NOAA Geophysical Fluid Dynamics Laboratory on Chapter 5; and Dr Stephan Harrison at the School of Geography and the Environment, Oxford University on Chapter 6. I am also very lucky as an Oxford resident to have been granted full access to the University's Bodleian Library, and am very grateful to staff there, particularly at the Radcliffe Science Library, for all their help and assistance with my research.

The manuscript was also reviewed by non-scientists

(though all experts in their own fields), who helped immensely with my laboured efforts to improve faltering first drafts of the book. My deepest thanks on this front go to Hugh Warwick, Anuradha Vittachi, Katharine Ainger, Greg Muttitt, George Marshall, Oliver Wright and Dorothy Atcheson. My friend George Monbiot even took on the onerous task of reading several parts of the book twice, and for his sage and refreshingly frank advice during the process I am particularly grateful. Another fellow writer, Paul Kingsnorth, who was working on his own excellent book *One No, Many Yeses* at the same time I was writing *High Tide*, shared with me the ups and downs of a process which was as rewarding as it was sometimes tortuous. Luckily we were able to give each other encouragement and support over many well-deserved pints in some of Oxford's best pubs, and I am especially thankful for his dedication in reviewing and commenting on so many different versions of my various chapters.

I should not forget my travelling companions: Karen Robinson, Franny Armstrong and Tim Helweg-Larsen, without whose contributions in both logistics and companionship my travels would have been much more limited and a lot less fun. Having been through some fairly epic adventures together, I hope we can all count ourselves lifelong friends.

I would also like to thank other friends for their help and encouragement over what has been a challenging three years: Emma Trickett, Sarah Hutcheson, Quentin Sommerville, Caroline Lowes, Al Chisholm, Lucy Ginsburg, Andrew

Wood, Caspar Henderson, Stephen Stafford, Lorna-Dawn Creanor, Angharad Jones, Jane Ritchie, John and Marion Manton, Helena Earnshaw, Jacklyn Sheedy, Will Ross, Eka Morgan, Jen Linton, Aubrey Meyer, Simon Retallack, Seb Naidoo, Joanna John, John Liverman, everyone from Seize the Day, Hugh and Betsy Howard (not forgetting Sarah and Elizabeth), Antonia Meszaros, Stuart Tanner and Anisha Charania.

Many different people offered invaluable assistance in the different locations that I visited. In Tuvalu, I was particularly grateful for the help I received from Kilifi O'Brien, Mataio Tekinene, Paani Laupepa and Ian Fry. Ove Hoegh-Guldberg's invitation to the Heron Island marine research station in Australia really opened my eyes to the severity of coral bleaching, whilst in Vanuatu Nelson Rarua took two days out to show me around the island and share several bowls of incredibly strong kava. Unfortunately space did not permit a full account of my very memorable stay there with him.

In Alaska, I also have to apologise to the residents of Arctic Village for not having the space to include more about their epic struggle to save the Arctic National Wildlife Refuge from oil drilling. During my visit there, Sarah James and her brother Gideon gave both time and a place to stay. In Shishmaref, Pastor Kathy Franzenburg likewise provided essential accommodation, and Robert and Jeanette Iyatunguk and Shirley and Clifford Weyiouanna helped with interviews, sage advice and marvellous sourdough pancakes. In Huslia, Cesa Sam was a godsend, and I am grateful to her brother Gabe Sam for putting us in touch. Wilson and Eleanor Sam made a huge impression on me with their quiet confidence and timeless wisdom, something which I will never forget. In Kaktovik, Nora Jane Burns and Robert Thompson gave invaluable help, and allowed us all to see a polar bear, and Ed Traynor got us a good price and a warm welcome at the Waldo Arms. None of my travels in Alaska would have been as fruitful had it not been for the quiet support of Larry Merculieff, who in the short time we spent together taught me more about life than perhaps anyone before or since.

In China, Professor Lu Qi is the unsung hero – it was his connections and organisational skills which enabled me to travel freely to Inner Mongolia and Gansu, and to meet Professor Liu Xinmin. Mr and Mrs Dong put me up for three days and fed me sumptuous meals in Yang Pangon village without a murmur, and would allow no payment by way of thanks except a box of airport chocolates. Most importantly, I would have been completely lost had it not been for the tireless efforts of my interpreter Liu Zexing. Any visitor to China would benefit from his help: contact www.chinatour-translator.com or richardliubeijing@ hotmail.com.

In mainland America I was lucky enough to be able to call on extended family for help and a place to stay – my great aunt and uncle Jean and Dick Atcheson put me up at short notice in both Princeton Junction and Washington DC, as well as offering useful expert advice on literary matters into the bargain. Their son and daughter, my cousins Nick and Katie Atcheson, also put me up and introduced me to the delights of the downtown clubbing scene in New York.

Finally, in Peru, Alcides Ames provided invaluable logistical help as well as offering his own expert knowledge on glaciology. His 'Mi Casa' hostel in Huaraz comes highly recommended for all visitors. Also in Peru, Benjamin Morales Arnao (who also knew my father), Marcos Zapata, Bryan Mark, Georg Kaser, William Tamayo and Bernard Pouyaud gave time and expert viewpoints which were invaluable in helping me understand the magnitude of the changes in Peru's glaciers. I am also truly grateful to my driver Maycoln Campos Urra (attached to the Hostal España in Lima) for getting me down from high altitude so quickly once things turned ugly, and recommend him to others who plan to get themselves into the same predicament.

This book might never have made the difficult transformation from idea to reality had it not been for two people: my agent Antony Harwood, who took me on and sold me relentlessly with little more to go on than a pitch at a party, and my editor Philip Gwyn Jones at Flamingo, who literally saved my proposal from the bin and surprised me with both a contract and an advance when I was just on the point of giving up. Their support has been unstinting, their contribution to this book inestimable, and they have my deepest thanks. Many thanks also to my editor at Picador in New York, Frances Coady, for taking the book on for an audience I was particularly excited to reach in the US.

The last word has to go to my family. My parents, Val and Bry Lynas, believed in me and this project with such unstinting loyalty that I can barely put my gratitude into words. Their love and support saw me through from the beginning, and it is their progressive and ecological world view that my own environmentalism grew out of in the first place. Likewise, my brother Richard and sisters Jenny and Suzanne helped in more ways than I can possibly acknowledge. Jenny and her husband Steve's young family, my niece Amy and nephew Thomas, also kept me mindful of what was perhaps a more profound motivation for undertaking this book. For it is Amy and Thomas's generation, more than mine, which will have to live through the full consequences of the lifestyles we are all leading today. It is their future which is truly at stake, and I hope this book can make some contribution, however small and intangible, to helping safeguard it.

Prologue

There was something different about the rain that night. I noticed it as I lay awake – a purposeful, remorseless drumming on the roof, as if determined to force its way into the house. It rained all night, the claustrophobic intensity of the downpour leaking insidiously into my dreams.

The next day dawned bright and even warm, its weak autumnal sunshine driving away the uneasiness of the previous night. But half a mile from my house, the Thames already looked very different. Instead of the usual sluggish flow, the brown water was racing angrily by. Small whirlpools and eddies played in the strong current, and freshly-torn branches floated past.

The wildlife too indicated that something was wrong: hundreds of earthworms, forced out of their holes by the water, were wriggling uselessly on the banks. Some of the lower watermeadows had been submerged by the rising river, and deep chalky puddles lined the towpath as I splashed through on my bike.

It was almost as a challenge to the elements that

I dragged a friend's kayak down to the riverbank, and – after a brief wobble of trepidation – launched myself into the water. The last thing I remember seeing, as I shot out into the strong current, was my bike propped against a willow tree on the bank.

As I sped downstream, it gradually dawned on me that I had made a mistake. I couldn't turn around without the risk of capsizing, and I didn't want to find myself flung into the water so close to the weir. I had already taken the right-hand fork at the island, under the 'Danger' sign that warned bigger boats away, and could hear the muffled roar of the rapids ahead.

A few minutes later I could even smell the spray. I eased closer to the bank as the weir came into view, its ugly steel gates fully raised to let through the maximum volume of the swollen river. On the right bank, under a grove of poplars, was the grey concrete memorial to an Oxford University canoeing team who had lost their lives in the same spot almost a century before.

As I should have known it would, the increasing current took me by surprise. I had aimed to pass opposite the weir by hugging the bank on the other side, and then continue on round to the main channel. But the tug was stronger now, and within a few seconds I was away from the bank and losing control. I tried to paddle backwards, but succeeded only in spinning round with a dangerous near-capsizing wobble. All the while the roaring waterfall moved inexorably closer.

I dug the paddles in deeply, as if on a liquid treadmill.

I pulled harder, gasping from both the exertion and the adrenalin, until the speed of my kayak began to gradually outpace the water flowing underneath.

Nearing the bank again, I grabbed at an overhanging willow branch. The whole thing snapped off – it's not called 'crack willow' for nothing – the surprise nearly catapulting me into the water. Instead I snatched at a handful of stinging nettles and thorny brambles, clinging to them with relief as I came in close to the bank and safety.

For weeks afterwards the placid Thames became virtually unnavigable. Within hours of my ill-considered kayaking trip, much of the Botley and Abingdon roads – two of the principal traffic arteries leading into Oxford – were underwater. It didn't spill over from the river directly, but instead appeared spontaneously in low points along the road, surging out from manhole covers and drains. In places the water was over a foot deep.

Further towards the centre of town, Osney Island was also inundated: each house had sandbags across the front door, and small rivers were beginning to flow down some of the lower streets. On the other side of the road allotments were gradually disappearing under the muddy flood. Sightseers were savouring the unusual scene.

A white-haired woman appeared on the towpath. 'It's all in the Bible,' she told me furtively, before scampering back behind her garden wall. 'This is the beginning of the end.'

She wasn't the only one to feel something different in the air, that week back in autumn 2000. Radio phone-in shows echoed with a phrase that had been in the national subconscious for a long time, but which was now breaking out into the open: *climate change*. Long the preserve of only scientists and environmental campaigners, the phrase began to reverberate in day-to-day conversations across the country. The British have always talked about the weather, but the weather was no longer behaving like it used to. Something was wrong.

Even politicians showed signs of noticing it. Tony Blair flew to flooded areas in a helicopter, announcing to journalists his determination not just to improve flood defences but to 'tackle climate change at the international level', whatever that meant. Blair's deputy John Prescott paraded around in wellington boots, also looking suitably grim for the cameras. 'All these incidents of climate change are reminding everyone, wherever they are in developed or developing countries, that this affects us all,' he lectured sternly, hiding from the bucketing rain under a temporary shelter of reporters' umbrellas.

As the rain poured down, the political climate was changing too.

Not being a scientist, I didn't know much about global warming, but I did know some of the basic facts. I knew that the Earth had warmed by over half a degree centigrade during the twentieth century. I knew that the rate of warming had doubled since the 1970s. And I knew that eight of the warmest years on record had occurred since 1990.

I understood the underlying science which explained

why this was happening: that every year six billion tonnes of the 'greenhouse gas' carbon dioxide pour into the atmosphere, and that this comes from familiar sources like car exhausts, power-station chimneys, domestic boilers and the destruction of forests. I knew that levels of carbon dioxide in the atmosphere had risen by a third since the start of the Industrial Revolution, and that those of another greenhouse gas, methane, had doubled. The science explained that these gases acted like the glass on a greenhouse, preventing the sun's heat from radiating back into space.

But it was all a bit too abstract, and I found it difficult to connect to my everyday reality. Was this why I hadn't seen a decent snowfall in years? Did it explain the floods too? Was it the reason why the weather was suddenly so hot that spring?

As part of my work at OneWorld.net – where I spent five years – I had covered Hurricane Mitch in Central America, droughts and famines in Africa and Asia, the catastrophic floods in Mozambique and the killer mudslides in Venezuela. Did all of this also connect up into a bigger picture of global warming? I suspected it might, but it was only a suspicion, more of an intuition than a conclusion based on any firm evidence.

Although I wasn't sure how climate change might fit into it, my concern about the environment was a long-standing preoccupation. I grew up in a country being rapidly destroyed by economic 'growth' and ever-increasing consumption. I saw my local shops in Oxford struggling and then closing down once big supermarkets opened on the outskirts of town. I saw pubs with real heritage turned into heritage theme pubs, and streets turned into elongated car parks. I saw rat runs and road rage, and I coughed in the fumes of my own car in motorway traffic jams on the way to work.

Then, spurred into action like many others during the mid-1990s, I climbed trees and dug tunnels to try and stop the building of yet another road. I loved the tall trees, the clear streams and the shaded wild garlic meadows, and something inside me snapped every time I heard the chainsaws and saw the old oaks and beeches come crashing down. I'd grown up trusting that things would generally get better, and that 'progress' worked. Now I wasn't so sure. Even in itself the destruction seemed senseless – but what if it was presaging a wider destruction, giving birth to a society which would poison itself for decades into the future as well?

After the Newbury Bypass was completed, I drove through it, trying to recognise landmarks on either side of the open concrete wound. I saw it fill up with traffic, every car pumping more greenhouse gases into the atmosphere. And I joined the activists' reunion on the day we blocked the road with a banner and hurled stones down onto the crisp new tarmac from the lifeless earthworks all around.

But what if I'd been missing the bigger picture? What if the real Newbury Bypass tragedy wasn't just that a forest in Berkshire was destroyed, but that an island in the Pacific was drowned or a Nicaraguan family swept away in a flash flood? What if the local was the global after all? My work at

OneWorld had given me an insight into the wider world, and I felt the connection, but it was a hazy picture.

It wasn't long before I would start seeking answers. My journey had almost begun.

I spent that Christmas at my parents' small farm in Llangybi, North Wales. It was still raining, and I had to complete the trip up the coast from Machynlleth by bus, because the railway track was cut by flooding (an increasingly frequent occurrence, according to the bus driver). On Christmas Eve my father and I spent the day stretching new fencing around one of the sheep fields, stabbing the rocky earth with a long iron bar to make holes for the wooden posts.

On clear days there is a glorious panorama from Snowdon in the north down to Cadair Idris and even Pembrokeshire in the south, but we could see nothing, just mist and relentless soaking drizzle. Even the sheep looked miserable and bedraggled as they nibbled at the close-cropped grass. We worked almost in silence, pulling out muddy stones by hand and banging the posts in quickly before the narrow holes filled with water from the saturated ground.

My mother had a fire blazing in the living-room stove when we came back in, and with my brother and sisters also having arrived we sat down for the customary family slideshow. My father set up the projector, whilst I moved a painting to make enough white space for the image. He'd selected slides of Peru, where the whole family had lived during his overseas geological posting between 1979 and 1982.

The projector whirred, and there we all were, back in 1980 – my brother Richard, just a few weeks old, yelling in a pram outside the house; my little sister Suzanne looking startled in a flowery dress; my elder sister and I naked and tanned on a dark-sanded Pacific beach, building sandcastles; then all of us in the Peruvian Andes, the snowy peaks of the Cordillera Blanca towering behind us. I'd seen the pictures dozens of times, but I was still entranced. My father flicked on to a field trip he'd done the same year with his geologist colleagues into a place called Jacabamba.

'That's my altitude record,' he commented gruffly, over a slide of a pristine snowfield, gleaming brilliant white in the tropical sun. 'Five thousand, two hundred metres.' The projector whirred again, revealing an enormous fan-shaped glacier looming over a small lake. Icebergs were floating in it, having tumbled down from the glacier above. It was a spectacular sight.

'What a place!' I breathed.

'I loved it. Incredibly hard work, carrying drilling equipment around to take rock samples, and then spending freezing nights in those crappy old orange tents. But I loved it.'

'It may not be the same now. I've heard that glacial retreat in the tropical Andes is pretty rapid.'

'Perhaps. But that was a pretty big glacier. Once an iceberg that calved off it into the lake created a huge set of waves which washed away half our equipment. There were

ice avalanches coming down from above the whole time.'
He paused. 'Still, maybe it has changed. I don't suppose I'll
ever go back, but I wonder what it *does* look like now.'

'I wonder . . .' I repeated slowly.

Then I said nothing for a while. I'd just had an idea.

That night, in many ways, marked the start of the journey
described in this book. Over the next three years I would
visit five continents, searching for the fingerprints of global
warming. I would interview Mongolian herders, Alaskan
Eskimos, Tuvaluan fishermen, American hurricane chasers
and a whole army of scientists, all with an urgent story
to tell. It was a story, when I first heard it, that left me
both shocked and inspired: shocked, because of how few
outsiders realised the magnitude of what was unfolding,
and inspired because there was still time to avert a far
greater catastrophe.

This global quest wouldn't be easy, and at one point it
would almost cost me my life. But at the end of it all I knew
I would return to Wales with a box of slides. I would draw
the curtains. I would set up the projector. And then I would
answer my father's question.

Britain's Wet Season

It was still raining, and York station was in complete chaos. The railway track was underwater both north and east of the city, and trains for Edinburgh, Newcastle and Aberdeen were terminating there, disgorging their tired and confused passengers into the mêlée. People dragged their luggage in and out of crowded waiting rooms as train after train listed on the display board was cancelled. Harassed staff tried to show passengers alternative routes via local buses, whilst others simply fled from the station concourse, pursued by angry travellers demanding to know how they could ever reach their destinations.

It was the beginning of November 2000. By the end of the month Britain would have experienced some of the heaviest rainfall and worst flooding ever recorded. On the miserable Friday night I arrived in York, newspapers and radio shows were already buzzing with speculation. This wasn't normal, everyone agreed. Floods had come and gone before, and Britain was supposed to be rainy. But no one could remember anything like this. There had to be some new explanation.

October had also been a washout. On 11 October Kent, Sussex and Hampshire received ten centimetres of rain – more than a month's usual average in a single day. Sixty government flood warnings were issued for the southeast of England, and the residents of Uckfield awoke to find their town centre under more than a metre of water. Lifeboats rescued people stranded in their homes, and one shop-keeper was washed away by the rising flood as he tried to open his shop door. Horrified neighbours looked on as he was sucked down the high street by the torrent. 'He didn't even have a chance to scream, the water was so fierce,' one told a *Guardian* journalist. ('"Unheard of" rain sweeps the south' was the newspaper's dramatic head-line.[1]) Happily, the shopkeeper was later found clinging to a riverbank.

Close by, a supermarket's windows caved in under the pressure of the water, and stock began to float off the shelves and away down the street. In Lewes, a town down-stream on the same river, council staff had to drive around with a loudhailer warning residents in low-lying areas to evacuate to higher ground. Six lifeboatmen were lucky to escape with their lives when their boat was nearly trapped under a bridge.

And still it kept on raining. The government's country-side minister, Elliot Morley, was one of the first to acknowl-edge something unusual when he visited the area the next day. 'We seem to be having more violent weather patterns and we accept that it could be due to global warming,' he said.[2]

Was the minister right? Had climate change indeed come home to Britain?

York was dark and eerily deserted. The heavy rain had turned to heavy sleet, and just a few cars splashed through the huge puddles that had gathered in the road. I walked along beside the old city walls, down towards the river.

The Ouse was almost unrecognisable. There was no sign of a riverbank – instead the water reached right round the buildings on both sides, and was almost touching the top of the arches of the road bridge. In the glow of the street-lights it looked as slick as oil, but also seemed to be moving impossibly fast, swirling forcefully around the stones of the bridge. In both directions streets which had usually led to boatyards, pubs and restaurants were deserted, the bustle of people replaced with lapping black water.

The worst of the rain had fallen two days earlier, when an intense depression – the remnants of an Atlantic hurri-cane – crossed the country, dumping several inches on the Pennine Hills. With the ground already saturated from previous deluges, the new water simply sluiced off into the rivers Nidd, Wharfe and Aire. The Aire valley was par-ticularly hard-hit, and in the Yorkshire towns of Keighley, Skipton and Bingley families had been forced to camp out in leisure centres and bed and breakfast accommoda-tion. Further downstream in Leeds the runoff overtopped embankments usually eight metres above the water level, turning city streets into canals temporarily reminiscent of Venice.

York is often hit by floods, but it was soon apparent that this disaster was off the usual scale. The day before I arrived, the Archbishop of York had paddled around his palace in a dinghy, whilst tourist rowing boats had been commandeered to evacuate an old people's home. That day the water was within half a metre of breaching flood defences, which would have submerged another seven hundred houses.

On 2 November, as I peered over the bridge at the rising River Ouse, the nationwide floods were already the most extensive on record. But the worst was yet to come.

No one knew where I could catch a bus to Scarborough (the railway line was under a metre and a half of water at Malton). I found the coach almost by accident on the station forecourt, already besieged by bedraggled travellers, most of whom wanted it to be going somewhere other than Scarborough. Rain was still coming down in torrents and people hurried on board, shaking off their umbrellas on the bus steps. The journey took much longer than usual, and as we passed through the Yorkshire lowlands the darkness outside was inky black through the steamed-up windows.

My sister's husband Steve was waiting in his car when we arrived.

'The main road to Filey's cut off,' he said. 'But there are other ways in on the country roads, so I'm pretty sure we'll get through.'

As we left Scarborough the rain turned halfway back into sleet, and began to pelt down at an even more incredible rate. Steve had to slow right down, and with the headlights on full beam the drops falling from the sky seemed to unfold like curtains. Water was simply sheeting off the fields into the road, collecting in any dips and low points in large ponds. We passed a big roundabout near Filey which was almost entirely submerged. In the flashing headlights of an emergency vehicle I could see a stranded car in what looked like a lake. We tried several other routes, before finally making it through to Filey on the last open road.

The scale of the damage became clear next morning. Just across the street from my sister's house is a small stream which runs down to the beach in a narrow cutting, next to a tarmac path which is shaded by trees in the summer. All the way down the valley the path had been ripped up – huge slabs of tarmac tossed around and dumped with piles of other debris on the beach. Rubbish was stranded a metre or so up the trees, showing how high the water level had reached. I hadn't seen it happen, but it was clear that what had taken place in that quiet valley was virtually a flash flood. Filey was still officially cut off, and all the way along the back of the beach mudslides had fallen from the saturated cliff face. In the town itself, various front gardens had turned into small lakes.

Even then it didn't stop raining. There was a brief respite for a couple of days, but weather reports identified another storm already gathering out in the Atlantic, where higher-than-average sea surface temperatures were giving the depressions more energy and moisture than usual. In York more than 4000 homes were evacuated as the river crested at levels unmatched in over a century. The tiny village of Naburn, just south of the town, became an island – veterans of the Mozambique floods from the International Rescue Corps were drafted in to help safeguard lives and belongings. Nationwide the death toll now stood at eight. In Naburn there was some good news: a baby was born, tended by a midwife also marooned on the new village-island, and the milkman was still able to make deliveries, albeit by Land-Rover.

On Sunday I returned to York – by road, since the railway was still cut off – to find a city under siege. The sense of crisis was heightened by crowds of sightseers, and by TV crews giving breathless live reports in front of the still-rising river. Brown water lapped around the foot of the castle mound, and in many riverfront streets pumps were fighting to extract water from basements, multiple hosepipes discharging back out into the road. On the west side of the river sandbags had been piled high along the tops of walls.

As I walked around, leaks were springing up everywhere, leading in some places to mini-waterfalls cascading down the bank of sandbags. On the other side of this fragile barrier thousands of tonnes of river water were perilously close to escape – a fact that seemed lost on all the people who, like myself, wandered around carelessly below river level to take pictures. Somewhere on the other side of the defences was a riverside park, though only trees and the tops of bushes testified to its continued existence.

York escaped complete disaster by a mere five centimetres that day, although a thousand properties had
already been submerged. It was the worst flooding in four
hundred years of records. The River Ouse had peaked five
and a half metres higher than normal, and the city centre
was only saved by the round-the-clock efforts of police,
soldiers and firemen shifting 65,000 sandbags to hold back
the water.

And York was far from alone – in Shropshire, Shrews-
bury was also being hit hard, as was Bewdley in
Worcestershire. Downstream from York a lake the size of
Windermere, the largest in the English Lake District, had
formed: as my train travelled south back down to Oxford,
passengers gawped at the new inland sea (complete with
large white-capped waves) which had obliterated fields for
miles on both sides of the raised track.

As people cast around for a cause, different theories were
advanced. Some blamed new housing developments on
river floodplains, whilst others claimed that new farming
practices meant that water ran off ploughed fields too
quickly. But although both may have played a role, they
were far from being the whole story. The truth was much
more straightforward: Britain had simply been deluged with
a staggering amount of rain.

More than three times the normal monthly average of
rain fell in parts of the southeast and Yorkshire. Whilst in
most places the deluges were judged to be a once in a century event, rainfall totals were sometimes so extreme that

they far outweighed previously-observed natural variability. Plumpton in East Sussex recorded 144 millimetres in a single twenty-four-hour period, something that would be expected only once every 300 years,[3] whilst the River Uck catchment in the same county had a thirty-day rainfall total that should only occur once every 650 years.[4]

Most floods come and are gone again in a few days at most. But in October and November 2000 storm relentlessly followed storm, leaving no time for the water to drain away. In England and Wales the September to December 2000 rainfall total was the highest since records began in 1766. In major river basins the floods were the most extreme of any ninety-day period on record, and for shorter time periods were only outranked by the March 1947 'Great Floods' – which had been generated by rapid snow-melt and rain running off still-frozen ground, and thus were much briefer.[5]

But even very extreme events – which happen only once in a lifetime or even less – can still be part of the natural variability of the climate. A single flood, however dramatic and destructive, isn't enough to convince a scientist that global warming is to blame. In order to be able to identify more clearly humanity's telltale fingerprint on the climate, there has to be a *trend* – evidence of a definite longer and wider change for which other causal factors can be confidently ruled out.

THE 'SMOKING GUN'?

As it happens, such evidence is indeed available for the UK. To find out more, I went to visit the climatologist Dr Tim Osborn at the Climatic Research Unit (CRU), part of the University of East Anglia. It was almost two years to the day after the start of the autumn 2000 disaster, and there was a hint of winter in the air as the London train sped through the Essex town of Colchester, past the River Stour saltmarshes and on through the flat Norfolk fenlands to Norwich.

As so often, Osborn confounded my expectations. No white lab coat for him: instead, a youngish, fair-haired man in shorts, trainers and a red golfing T-shirt was leaning over the balcony three floors above as I arrived at the round, glass-fronted CRU building.

'Hello! Come on up,' he shouted as I climbed the stairs. His room was strewn with back copies of the *International Journal of Climatology* and meteorology books, as well as sheafs of paper – many covered with impenetrable algebraic scribbles.

'Sorry about the mess,' he said as I sat down on a free chair. Then he swivelled his own chair round to face the computer screen. 'Now, have a look at this.'

Osborn has spent years analysing nearly half a century of rainfall statistics. From a damp day in 1960s Blackpool to a torrential summer downpour in 1990s Devon, all these records were fed into his number-crunching computer. When spat out the other end into a series of graphs, these statistics – rather than just showing the usual random vagaries of the British weather – showed that something very unusual *was* going on. In fact the trend was so clear that even Osborn himself was 'surprised' by what it revealed.

What Osborn discovered was that over recent decades heavy winter downpours have indeed increased dramatically. 'Over the period from the 1960s to the mid-1990s there was a doubling of the amount of rain that came in the "heavy category" in winter,' he explained. 'So in the 1960s something like seven or eight per cent of each winter's rainfall came from what we call the "heavy" events, whilst by the mid-1990s that had increased to about fifteen per cent.'[6]

With more rain falling in a short time, river systems were unable to cope – and floods were the inevitable result. What's more, this heavier winter rainfall was directly related to rising atmospheric temperatures.

Straightforward atmospheric physics suggests this could be the global warming 'smoking gun'. The relationship between temperature and the air's capacity to hold water vapour is not linear – in fact the air can hold proportionally more water as temperature rises.[7] So in a given 'precipitation event', whether it is snow, hail or rain, more water is available to fall out of the sky over the same short period of time.

This is exactly what seems to be happening in Britain: as a result of global warming, more warm, saturated air rushes in from the Atlantic, causing stronger storms and heavier rainfall. As a result, the probability of heavy rainfall has doubled over the last thirty-five to forty years in southeast England, according to observations and analysis conducted by Osborn and his CRU colleague Mike Hulme.[8]

These aren't one-off downpours, either. The frequency of prolonged five-day heavy rainfall events has also been increasing. In Scotland floods have been getting far more frequent over the last few decades, whilst in England and Wales there have been four major floods in the last twelve winters: 1989/90, 1993/94, 1994/95 and, of course, 2000/01.[9] The match for 2000 isn't perfect because the worst flooding came during the autumn – but the floods also lasted right through until January, just as the trend would suggest.

Osborn's work also coincides with evidence from other parts of the world. Study after study has come to the same conclusion: that throughout Earth's mid-latitudes, rainfall is getting heavier and more destructive. There has been a steadily increasing rainfall trend in the United States through the twentieth century, and much of that increase has come in the heaviest downpours. A number of catastrophic floods in recent years – most notably the Mississippi floods of 1993, the New England floods of 1997 and the winter floods of 1997 in the Pacific northwest and California – seem to show the shape of things to come.[10]

Scientists have reached a similar conclusion in Europe,[11] whilst in Australia rainfall totals are also rising steadily.[12] This might seem to be a good thing in a continent often afflicted by drought – but again, much of the increase has come in the heaviest deluges, which are less likely to soak productively into farmland, and more likely to run quickly off the land in destructive torrents, taking the fertile topsoil with them.

One study looking specifically at large river basins – such as the Yangtze in China and the Danube in Europe – confirms what many people have long suspected: that big floods are indeed getting more frequent. In fact, sixteen of twenty-one 'great floods' during the twentieth century have occurred since 1953, and in the planetary mid-latitudes seven out of eight have also occurred in the second half of the century.[13] UK-based researchers have also identified a near-global trend towards heavier rain and floods.[14]

In the most comprehensive survey of all, the Intergovernmental Panel on Climate Change (IPCC) confirmed that rainfall was getting heavier and more extreme in the United States, Canada, Switzerland, Japan, the UK, Norway, South Africa, northeast Brazil and the former USSR.[15]

This hasn't affected everywhere: some places have got drier, such as the Sahel in Africa and northern China. But almost the whole of the Earth's mid-latitudes has been affected, and as Osborn told me, 'if there's something coherent going on at all the mid-latitudes, then there must be something virtually global scale driving it'. Computer models of global warming have long illustrated this effect, and now it seems to be showing up in the real world, just as many scientists – including Osborn himself – have long predicted.

MONMOUTH, FEBRUARY 2002

Just under 10,000 homes were flooded in Britain during the 2000 event. Some were hit two or three times, and a few left completely uninhabitable. Transport and power services were disrupted, and the cost of flood-related damage eventually totalled around £1 billion, according to the government's Environment Agency.[16] Everybody breathed a sigh of relief when it was finally over, but only one year and three months later – in the first week of February 2002 – the floods were back.

This time one of the worst-hit places was Monmouth, a historic town just over the Welsh border at the confluence of the rivers Monnow and Wye. On 4 February 'severe flood warnings' were issued for both rivers, schools were closed and residents in low-lying areas began to move themselves upstairs. Twenty families were evacuated from mobile homes when the Wye burst its banks, and three streets were completely submerged.

Judging from the news Monmouth sounded well worth a visit. This meant hiring a car, but I was ready to leave by mid-morning, heading towards Cheltenham on the old A40. The River Thames was pretty high, and when the road crossed some small rivers on the way over the Cotswold hills, I could see that each was swollen, its banks only identifiable by lines of willow trees standing in the brown water.

Just outside Gloucester was the first sign of large-scale flooding – a huge new lake stretched almost as far as the eye could see. Trees, telegraph poles and even an electricity substation were surrounded by water, and a couple of swans paddled by.

I drove on. The sky was darkening again with ominous clouds as I neared Wales, and soon a heavy shower sent torrents of new water coursing down off the hillsides.

About ten miles outside Monmouth I spotted a 'Road Closed' sign and drove round it to investigate. I was deep in the Forest of Dean, and the small road led down a steep wooded valley towards the River Wye. On the river itself was a small village, little more than a hamlet, called Lower Lydbrook.

Lower Lydbrook looked like it had been doused entirely in mud. Mud was everywhere: across the road, the pavements, people's drives and lawns. The whole area had clearly been awash with very dirty floodwater only a few hours beforehand. Outside the Courtfield Arms a man was sweeping the sticky brown mess off the car park. I slithered up to him and asked whether he felt the flooding was getting worse.

His answer was surprising. In the past the floods had come once every three or four years. Now it was two or three times in a single year. And the latest inundation was easily the worst for three decades.

On the other side of the road was a restaurant called the Garden Café. All the gravel in its neat drive was coated with the same brown layer, as was a car parked outside. I followed some fresh footprints round to a side door. It was swinging open, and I peered into the gloom inside. Not surprisingly the place was a mess: fridges were stacked up on tables and wet rugs were hanging from the beams. There was a pervasive damp musty smell, and a clear high-water line about a metre up the walls.

The owner was happy to take a break from cleaning up, and introduced himself. 'Paul Hayes. Owner and chef of the Garden Café.' He looked around at the disastrous mess and added: 'Currently on holiday.'

Hayes was certain that the flooding had got worse in recent years. It wasn't necessarily that more rain fell overall – but rather than being averaged out over a month, the whole lot simply fell in one night.

'We don't have a winter any more, we have a wet season. It's like tropical rainstorms here. And because it's a hilly area this translates into flash floods. The river rose six metres from its level last week. It came in here at four on Sunday morning, and within another two hours reached a metre up the wall. It never used to flood in the house, but that's three years in a row we've been flooded now.'

As a result, his business was wrecked. All the fridges were ruined, he was losing customers every day the restaurant remained closed, and all his stock would have to be thrown away. Nor was this the first time: during the winter of 2000 – when the building had been flooded on three separate occasions in October, December and January – he had only managed to open for twelve days throughout the whole four-month period. And with the whole property now virtually uninsurable, no buyer would even look at it.

Hayes had a knowing, worldly manner, but I could tell that even he had been thrown by the latest deluge. 'It came so suddenly,' he said, almost perplexed. 'I knew it was going to flood, even though there was no flood warning. And if it rains in the next week it'll flood again – all that water's got nowhere to go.'

In Monmouth itself the floodwaters had only just begun to recede. Most of the town was unaffected – the Romans had sensibly founded it on a hill, but developments in more recent centuries have extended the town right along the river. Built at the confluence of two rivers, and not far from the tidal estuary, the area has always been prone to flooding – the one reliable crossing point has been called Dry Bridge Street since Norman times.

When I arrived, though, Dry Bridge Street was half underwater.

Children were splashing around and riding their bikes through it, whilst dog-walkers in wellington boots waded through to a nearby park. Sandbags were stacked in front of every front door. Opposite the bridge itself, the Green Dragon pub had narrowly missed inundation just hours earlier. A hundred yards away, the Britannia Inn had not been so lucky, and water was still being pumped out of it into the road.

I knocked on the door and it was opened by a young woman with short brown hair.

'We're closed because of the floods,' she began, looking at me as if I were stupid. But when I explained what I wanted, she invited me inside.

Several regulars were sitting on benches reading newspapers in the gloomy half-light. A couple of others were helping sweep mud off the stone floor. Everyone agreed that the flooding was getting worse.

'This place is rotting,' complained the landlady. 'There is constant damp from the rain and sewage.' She poked disapprovingly at some blistered paint on the lower walls. 'It just keeps getting flooded. In the past it didn't seem as often – now it's twice a year. It's just constantly, all the time. It's hard enough to make a living in this trade as it is, without all this happening.'

'Thirty years ago you knew what the seasons were,' one of the regulars added, leaning on his broom. 'Now you don't know. It's got to be to do with the way the weather changes – the rainfall is unbelievable.'

I drove out of Monmouth and into Wales, the first mountains rising up in the distance. It was raining again, and just before Crickhowell flood warnings appeared by the side of the road. A small house next to a layby was completely surrounded, the water so deep in places that only the tops of the roadsigns stood out. I reached Machynlleth and my old friend Helena's house, on the west coast of Wales, long after dark, and lay awake listening to the rain hammering on the roof long into the night.

Machynlleth has a small museum-cum-art gallery called the Tabernacle, a compact slate-roofed building not far from the railway station. I headed down there in the morning with Helena. Not being a huge fan of the abstract oil paintings on the wall, I tried instead to engage the white-haired old lady behind the front desk in conversation. It's always easy, whether you're in England, Scotland or Wales, to strike up a conversation about the weather.

'Terrible weather, isn't it?' I ventured. The old lady carried on arranging some leaflets on the desk. I noticed her hearing aid, and tried again, more loudly.

'TERRIBLE WEATHER, ISN'T IT?'

'Oh yes,' she answered, 'such a lot of rain.'

I nodded encouragingly, and she went on. 'The last few years we've had more rain than I ever remember.' She paused. 'And no snow either. The last *proper* snow,' (and she emphasised the word 'proper' to show that she meant snowploughs, the town cut off and so on) 'was over twenty years ago. The snow we've had in the last few years has been hardly anything. Instead, it's been rain, rain, rain.'

On sale next to the desk were several Christmas cards, each showing children making a snowman under a heavy winter sky, the pretty white flakes swirling around them as they gathered up the snow in their duffle coats and woolly mittens. It was the traditional British winter, everyone's dream of a white Christmas. And what no one knows – or likes to admit – is that it's probably gone for good.

SNOW PLACE TO GO

Snow was becoming a rarity even during my childhood. Apart from the years in Peru, I grew up in a small Nottinghamshire village called Colston Bassett – a tiny place with little more than a pub, a primary school, and a local dairy famous for its pungent stilton cheese. Every autumn the village held a harvest festival, when all the local farmers would bring their produce into the village hall for a lavish evening meal. I looked forward to it for two reasons: because I and the other village kids were allowed to get drunk on cider; and because it meant the onset of winter.

I loved winter. From the first frosts in October to the bursting of the buds in April I'd scan the skyline almost hourly for snow. It came, too: we even got snow on Easter Sunday one year. In January 1987 it fell so heavily over-night that the drifts piled up against the side of the house and meant a day off school. The school bus got through after a couple of days, but the snow lasted for almost a fort-night. Every winter there'd always be a few centimetres of snow which would generally last for two or three days. I was filled with barely-suppressed excitement each time the first flakes fluttered past the school windows.

I haven't seen snow like this for over seven years in Oxford, which isn't too far from where I grew up. Back in 1996 there were a few days of snow (no big deal, less than ten centimetres deep. I remember it principally because I fell off my bicycle on the ice) but since then nothing. In fact snow has become so rare that when it does fall – often just for a few hours – everything grinds to a halt. In early 2003 a 'mighty' five-centimetre snowfall in southeast England caused such severe traffic jams that many motorists had to stay in their cars overnight. Today's kids are missing out: I haven't seen a snowball fight in years, and I can't even remember the last time I saw a snowman.

A quick glance at the official weather records for Oxford

confirms my rather hazy impressions. The last decent snow was in 1985, when there were twenty-one days of snow cover. The winter of 1963 was the most extreme in England since 1740, and during the 1970s snow days averaged about eight days per season. How things have changed. Six out of the last ten years have been completely snowless, whereas between 1960 and 1990 there were only two snowless winters during the whole three decades.[17]

By the 2080s our grandchildren will only experience snow on the highest mountaintops in Scotland, because over most of the English lowlands and the south coast snowfall will be virtually unknown.[18]

Other familiar things may also look very different. Take the average British garden. Lawns will need mowing all year round, and will die in summer droughts unless heavily watered. Traditional herbaceous border species like aster, delphiniums and lupins will also struggle in the dry soils. Tree-ferns, palms, bamboos and bananas will replace holly, oak and ash. Many fruiting trees and bushes need winter chilling for bud formation, so blackcurrants and apples will need to be replaced with peaches and grapes. Over-wintering bulbs need low temperatures to stimulate their development, so gardeners will need to dig up the bulbs and refrigerate them for a few days in order to coax spring flowers out of them. New pests and diseases will spread out of the greenhouse and into the open garden. Aphids, for example, begin their infestations two weeks earlier for every 1°C rise in temperature.[19]

Many of these changes are already underway, but have

been accelerating over the last two decades. Termites have already moved into southern England. Garden centres are beginning to stock exotic sub-tropical species, which only a few years ago would have been killed off by winter.[20] In Surrey, horse chestnut trees now come into leaf twelve days earlier than they did in the 1980s. Oak is coming out ten days earlier, and ash six days earlier. Winter aconites are now flowering a month earlier than three decades ago, and crocuses – which used to flower in March – are now putting out petals in mid-January.[21] The average UK growing season is now longer than at any time since records began in 1772. In 2000 there was hardly any cold weather at all: the growing season extended from 29 January to 21 December, leaving just thirty-nine days of winter.[22]

In the summer of 2003 temperatures broke through the crucial 100°F level for the first time in recorded history, peaking at 100.6°F (38.1°C) on 10 August at Gravesend in Kent. Continental Europe, meanwhile, suffered its highest temperatures for 500 years, sparking catastrophic forest fires in France, Spain and Portugal, and killing thousands of elderly people in the sweltering cities. In France alone almost 15,000 people died in the heatwave, sparking a national crisis of guilt and soul-searching as the bodies piled up. Even in the cooler UK, 2000 people died.

Heatwaves catch the headlines, but the insidious effect of higher average temperatures is having a permanent effect on our surroundings. Indeed, the temperature rise is now so rapid that in climatic terms English gardens are moving south by twenty metres each day.[23] (This is because, with every 1°C rise in temperature, climatic zones move 150 kilometres north.) English temperatures are predicted to soar by up to 5°C this century alone,[24] so by the 2080s our gardens will – metaphorically speaking – be nearing the south of France.

This is particularly bad news for 'heritage gardens'. The National Trust will be faced with the choice of uprooting everything from its much-loved English country gardens and trucking them to the north of Scotland, or giving up and letting the traditional species die.

In fact the British countryside our grandchildren grow up in is likely to be a very different place to the one we see today. According to the Woodland Trust, increased drought and water stress from hotter, dryer summers means that parts of London, East Anglia and the Midlands might become unsuitable for beech trees in the near future. Although beech woods on chalk soils should fare better (plant roots seem to be able to draw water large distances up through porous chalk rock), die-back has already begun in parts of East Anglia and Southern England. 'In the worst-case scenarios, beech could soon be absent from large areas of the south,' the Trust concludes.[25]

Oaks are also going to be on the endangered list. Although more likely to withstand summer droughts and winter floods, oak trees are threatened by a new disease called oak wilt – which has already devastated woodland in North America.[26] Oak wilt thrives in warmer winters: it could turn into a plague of similar proportions to Dutch Elm Disease, which virtually wiped out elms in the UK, once a common wood and hedgerow tree species. Because of Dutch Elm Disease, I have never seen a fully-grown elm tree: and when I was growing up every field boundary was lined with their enormous skeletal carcasses. Could oaks go the same way?

Instead of these familiar trees, woodlands are likely to be predominantly composed of sycamore, with other invasive species like rhododendron and Japanese knotweed making up the undergrowth. The animals which currently fit into our woodland ecosystems will also disappear – wood-peckers, butterflies, frogs and toads – all will need to move to cooler climes or die.[27]

In theory woodland species could 'migrate' further into the north and west of the British Isles to keep pace with the shifting climatic zones. Many butterflies and birds are already doing this: the speckled wood butterfly has moved north by over a hundred kilometres in the last sixty years – and it's still lagging behind current rates of climate change.[28] The nuthatch, a colourful tree-dwelling bird, is now extending its range, and the reed warbler has begun for the first time to breed regularly in Scotland and Ireland.[29]

But whilst birds and butterflies are clearly fairly mobile, most tree species are not. At the end of the last Ice Age trees could colonise new areas at a speed of up to a kilo-metre a year, by spreading their seeds and gradually establishing new saplings. But projected warming rates will far outstrip this: climatic zones in the twenty-first century will be shifting north seven times faster than most plant species can follow them.[30]

There are also some serious practical reasons why natural ecosystems can't simply move with a shifting climate, such as cities, enormous dead zones of intensive farmland and major roads. The great crested newt, for example, couldn't move north even if it wanted to – it can't cross the M4 motorway.[31] Nor are my local beechwoods likely to be able to get round Birmingham and Manchester in their supposed long trek north.

Extinction is a certainty for highly-specialised plants and animals which already live in very restricted areas. Norwegian mugwort, a plant which lives only in the Arctic cold of the highest Scottish mountain summits, simply has nowhere higher to go. Nor has the capercaillie, the emblematic Scottish bird which lives in pinewoods and is similarly dependent on low temperatures for its survival. Also on the way out is the natterjack toad – which according to a government study is due to lose its 'climate space' as early as 2020, when the seasonal ponds it breeds in dry out. The mountain ringlet butterfly will lose its climate space by 2050, and it too is slated for extinction.[32]

As with the National Trust's gardens, climate change will ruin British nature conservation strategies, which are currently based around a patchwork of Sites of Special Scientific Interest and nature reserves. Almost all of these are adapted to specialised habitat – such as upland peat-bogs, chalk grasslands or lowland heaths – which depend on particular rocks, soil and topography and therefore, by definition, cannot be moved.

Ecosystems are incredibly complex, with many different species occupying their own niche in the food web. Once these begin to fracture, specialised species will die out in all but the most tiny remnant habitats, to be replaced by only a few highly-adaptable weeds. Biodiversity will decline as these adaptable species, many of them invasive introductions from other parts of the world, take over ever-larger areas of our outdoors.

The British countryside of 2080 is likely to be an eerie, unnerving place, with the same familiar rolling landscape supporting only a few very mobile – but strangely unfamiliar – plants and animals.

Like the Christmas snow, the holly and the ivy may soon be distant memories.

Yet none of this has to happen, or at least not to the extent I've outlined above. Some amount of warming is already inevitable, but whether it reaches the extremes described above depends on all of us – and the decisions we take about how to run our lives and our economy. It depends crucially on one thing, and one thing only: how much greenhouse gas we release into the atmosphere over the decades ahead.

On the way back from Wales, I got caught in a traffic jam on the M6 just outside Birmingham. This one was a monster. Three lanes of cars, vans and lorries were packed solid. The whole place stank of petrol and diesel fumes, aggravated by a few irritated motorists who revved their engines pointlessly. A few drivers even got out and stood next to their vehicles, glaring at everyone else, looking for someone to blame. No one spoke. There was none of the camaraderie you often get on a broken-down bus or a delayed train. This was an atomising, frustrating experience. We were all trapped like prisoners in our little metal boxes, and every one of us hated it.

Despite jams and congestion, road traffic in Britain is rising inexorably. Every year Britons spend more time and travel greater distances in their cars. An increasing number of short journeys – under two miles in length, which could easily be done on foot or by bicycle – are now done in cars. Road-traffic levels rose by a fifth between 1988 and 1998, and are predicted to rise by nearly another two-thirds by 2031. Journeys by bicycle, meanwhile, are at an all-time low.[33]

In many ways car use is a self-reinforcing process. When I was young most children used to walk to school or go by bus. Now – partly because of parental fears about busy roads – the 'school run' has become one of the biggest causes of urban congestion. It causes gridlock every morning around eight on many of the roads near where I live. It's a vicious circle: the more parents who take their kids to school by car, the more cars on the road and the more dangerous the roads become for everyone else, forcing still more parents to resort to their cars. And so it goes on.

Similarly, the growth of out-of-town shopping has encouraged car use, putting town centre shops out of business and reducing the places one can shop without going in the car still further. By building new roads and supporting the growth of supermarkets the government has made matters worse – but we've all been complicit in these destructive trends.

And most people with cars can scarcely envisage living any other way. When the RAC recently asked motorists if they agreed with the statement: 'I would find it very difficult to adjust my lifestyle to being without a car', 89% said that they did.[34]

Nor can those of us who have given up our cars – but still, like me, regularly travel by jet aircraft – afford to be smug. A single short-haul flight produces as much carbon dioxide as the average motorist gets through in a year. The flights I undertook to research this book directly produced over fifteen tonnes of carbon dioxide[35] – which is equivalent to about forty-five tonnes once the overall warming effect of aviation pollution is taken into account.[36] Many people who work for environmental organisations travel enormous distances by plane every year – each with similarly valid reasons for doing so as I felt I had. Speaking personally, the impact of these flights is so enormous that it wipes out all the other aspects of my relatively green lifestyle (no car, green electricity, local food and so on) and is equivalent to my total sustainable personal carbon budget for about twenty years.[37] Oh dear.

Although cars are a highly visible pollution source, and transport accounts for a third of the average person's greenhouse gas emissions, another third comes from the home – over 50% of this from space heating.[38] With some insulation, a new boiler and some double glazing, heating costs and the associated emissions could be reduced dramatically – yet most of us don't bother, or simply can't afford to do so. Over a tenth of British houses have no insulation at all,[39] and 20,000 to 40,000 people – mainly the old and vulnerable – die every year because of cold-related killers like hypothermia and pneumonia.[40] As well as reducing climate change, better housing and insulation would save lives.

The other third of the average person's emissions comes from everything else – food, services and other consumer products, all of which generate pollution in their manufacture and transport. Many people now eat food from all over the world without even knowing it: green beans from Kenya join Chilean apples and Brazilian chicken on the average British dinner table. All these products – especially fresh fruit and vegetables, which mainly come by air – generate huge pollution costs as they are transported. None of that, of course, appears on the label. Nor, incidentally, does it appear in government greenhouse gas statistics or the Kyoto Protocol, from which fuels used in international transport are excluded.

So, ultimately, the extent of climate change is up to us, and this uncertainty about how we're all going to behave feeds through into scientific projections about future warming. The UK Climate Impacts Programme doesn't make single predictions: its latest report talks about different 'emissions scenarios' which might unfold during this century. In a 'high-emissions scenario' for instance, once-in-ten-year summer heatwaves may reach a blistering 42°C, as compared to 35°C now – and 39°C in a 'low-emissions scenario'.[41]

Overall twenty-first-century warming in southern England in a high-emissions scenario is 5°C, far too high and rapid to allow beech trees and other species to adapt. The low-emissions scenario on the other hand envisages average warming reduced to as little as 1°C, still dangerous (and nearly double that so far experienced globally) but probably low enough for most of our familiar species to cope with.

The story is the same with floods. In a high-emissions scenario, the report envisages an increase in winter rainfall by a third, and a doubling of intense downpours. In a low-emissions scenario – with less greenhouse gases in the atmosphere and reduced global warming – winter rainfall might rise by only a twentieth.[42]

The choice, it seems, is ours.

YORK, FEBRUARY 2002

Less than a week after my trip to Monmouth – and over a year since the autumn 2000 floods – I was back in York. The Ouse had risen again – not on the same disastrous scale, but enough to wash over the towpaths, flood into the riverside car parks and surround the trees and parkland on either side.

I walked down from the railway station and onto the main road bridge, peering over once more at the swirling brown water beneath. On the far side was a sign advertising boat trips, and so I headed down the steep stone steps.

At the entrance to the boatyard was a chalked sign announcing: 'Closed due to flooding'. The water was lapping at the edge of the yard so I skirted around the inside, to where the owner, John Howard, was sitting on a wooden bench staring at the river, absent-mindedly stroking a large ginger cat. Behind him workmen in blue overalls teetered across a makeshift jetty onto one of the out-of-service big boats, which was moored securely to two trees.

He waved me across to the bench. I stroked the cat, which started to purr loudly, and asked Mr Howard how the floods had affected him.

'I lost all my income for November 2000. We basically had to shut down. We had about a foot of water in the house . . .' (he indicated behind me to a neat cottage attached to a two-storey office) '. . . and if it happens more regularly we'll have to consider raising the floors. But flooding is part of the business – we just have to work around it.'

I asked if he knew of other river experts I might be able to talk to, and he disappeared inside the office, reappearing with a handwritten telephone number for Laurie Dews, an old-time bargeman. 'Now, Mr Dews, there's nothing he doesn't know about the river.'

I phoned him up straightaway.

'Oh yes, I think the floods have got much worse these days,' he said in a gravelly Yorkshire accent. 'There's all this heavy rain comes straight down off the hills. That's a big change.'

I asked whether he'd be available to talk that afternoon. He paused. 'Hold on. I'll just check with the wife.'

Mrs Dews had just got back from her daily walk when I arrived in Selby an hour later. She seemed a little frail, but her husband was still sturdy, with strong features and a lively face.

'Now what was it you wanted to know?' he asked after we were installed in two living-room armchairs with tea and biscuits. I admired a framed golden wedding photograph, dated 1996. Now seventy-nine, Mr Dews had been retired for over ten years. His whole family had been bargemen, he told me, and he could trace the line right back to his own great-grandfather. They all worked barges up and down the Humber and Ouse rivers, loading the boats with oilseeds at Hull and hauling the cargo up to the cattle-feed mills at Selby. 'We always respected the water,' he recalled. 'We didn't do anything silly on it.'

He sighed. The barges are all gone now though, he told me. Some had been sold off for houseboats, others had moved elsewhere or been scrapped. Most of the feedstuff transportation now went by road in huge lorries.

It didn't take long for us to get onto the weather. 'There's more flooding now than there used to be,' Mr Dews told me. 'Now you're getting more rain and wind for sure. In my opinion there definitely seems to be more rain now.'

'And less snow,' added Mrs Dews from the sofa.

'Yes, there used to be more snow, but you don't get snow any more. You don't get frost much either.'

Laurie Dews could remember the Great Flood of 1947,

whose effects had been worse than the recent event – but only because in November 2000 army Chinook helicopters had been employed round the clock to ferry sandbags to vulnerable points around the town. 'That saved a lot of Selby.'

He had got to know almost every aspect of the river in his long career: its shallow and deep areas, and how tides could have an effect on flooding. That's where global warming also came in. 'If seas and oceans are higher, your rivers aren't going to go out as well. Global warming has an effect there. It's got to have, hasn't it?'

'We're all doomed!' Mrs Dews interjected again from the sofa, giving me a wink.

'Global warming is bothering everybody now – but what can you do?' Mr Dews went on. 'With the floods, storms and sea level rise everyone's getting more concerned.'

Mrs Dews smiled over at us both. Outside the rain had begun again, and I was in no particular hurry to leave.

'How about another cup of tea?' she asked.

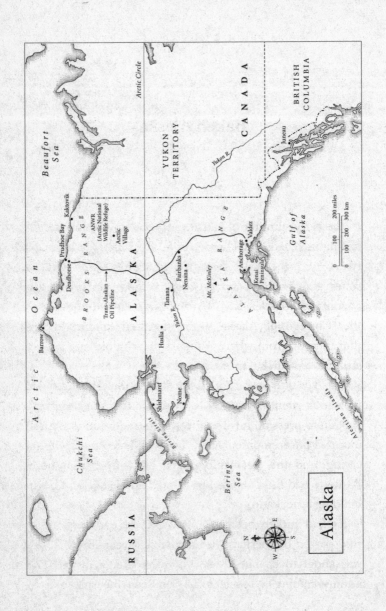

2

Baked Alaska

In the early summer of 1901 the steamship *Lavelle Young* –
having travelled 1500 kilometres up the Yukon and Tanana
rivers into the wild interior of Alaska – hit shallow
water and began to scrape the riverbed. On board was
an ambitious trader, E. T. Barnette, who hoped to establish
the 'Chicago of Alaska' three hundred kilometres further upriver, from where rumours of rich gold and copper
strikes were quickly spreading.

But Barnette never got to his destination. Having failed
(in a loud shouting match) to persuade the *Lavelle Young*'s
captain to press on further, the young trader found himself
dumped unceremoniously on the riverbank. As his wife sat
crying, and the steamship pulled away from the shore,
Barnette had little choice but to pick up his axe and begin
building a stockade.

He had initially been aiming to leave as soon as another
ship passed by, perhaps in as little as a year. But his luck
was about to change. A few months later a ragged and
hungry mining prospector, having seen the smoke from

Barnette's cabin, pounded on the door and announced that he'd just found gold.

Barnette decided to stay put and operate a trading post. And within two years his accidental settlement had become the largest log-cabin town in the world, with four hotels, two stores, a newspaper, a row of waterfront saloon bars and a thriving red-light district. Fairbanks had been born.[1]

Today the city still retains a frontier feel. Although the old log cabins now rub shoulders with shopping malls and fast-food outlets, moose still graze beside the busy dual carriageways, and bears roam the lowland spruce forests that surround the city for hundreds of kilometres. On the southern horizon stands the snow-covered Alaska Range, and on a good day you can see Mount McKinley itself, North America's highest mountain, gleaming in the far distance.

Fairbanks has always been a boom-and-bust town. In 1920, following the end of the gold rush, the town's population had dwindled to a thousand – after a high of nearly twenty thousand a decade before. Another boom came during World War II, when several large military bases were established to counter the Japanese threat. The army and airforce stayed on during the Cold War, and Fairbanks began to prosper as a military town.

But the biggest boom of all has proved to be oil. The Trans-Alaska Pipeline runs right past Fairbanks, and the town was the centre of the pipeline-construction effort in the mid-1970s. Much of the entire state's economy, not to mention its politics, revolves around the oil industry.

However, Alaska's current prosperity has come at a high price. Although few in the state care to recognise it, Alaskan oil – of which more than a million barrels a day are exported to the mainland US – has rebounded heavily on the state through global climate change. And whatever their views on global warming, almost every resident will admit one thing: Alaska's weather has gone crazy.

I arrived in Fairbanks late one evening after a twelve-hour train journey north from Anchorage. Most of it had been through the Alaska Range mountains, which were brilliant white against the blue sky, their tree-clad lower slopes speckled and green as the snows gave way to forest.

Travelling with me were Franny Armstrong, a film-maker, and photographer Karen Robinson. Franny filmed out of the doorway as the train rattled over deep gorges and through metre-high snowdrifts, whilst Karen snapped shots of snow-bound shacks buried in remote backwoods territory. Everyone gathered at the window as we passed Mount McKinley, but we were disappointed: the great mountain was hidden from view by grey cloud.

Once in Fairbanks we bundled our gear into a taxi and found a cheap backstreet hostel in a part of town where the front gardens were full of junk, and savage-looking dogs barked from behind chain-link fences. The hostel, which had several semi-permanent unemployed residents, was once a brothel, its proprietor confided soon after we arrived. This gave the otherwise unremarkable two-storey wooden building an air of seedy glamour, especially since the taxi driver had known exactly where it was, even calling it by its former name – 'Ruthie's'.

The proprietor, a young hunting enthusiast called Dale Curtis, watched us unpack.

'You guys tourists or something?' He adjusted his base-ball cap uneasily, leaning against the doorframe. (The doors were concertina cardboard, another brothel legacy.)

'No, we're journalists. We're investigating climate change.'

He looked blank.

'Global warming,' I continued. 'Asking people how the weather has changed and that sort of thing.'

He looked intrigued. 'Well, the weather sure has got strange. It don't get cold enough fast like it used to, and then it warms up real quick.'

This sounded interesting. I sat up and listened, encouraging him to continue.

'What really struck me was watching ducks swimming on the river all winter. It was Christmas time, January even, and they were still swimming around. They're not supposed to be here at that time, they're supposed to be south already.'

He shook his head in amazement, warming to the theme. 'And the bears come out too early. They don't know whether to go into hibernation or to wake up. Folks round here are real worried about it. A couple of years ago at Christmas it rained and melted all the snow away. That just ain't right, you know?'

As Dale Curtis was suggesting, Fairbanks is supposed to get cold in winter – really cold. Just a hundred and fifty kilometres shy of the Arctic Circle, in mid-December the town receives only three hours of sunlight. As any resident will tell you, the sun doesn't really come up at all – it just skirts along the horizon, as if entangled in the icy peaks of the Alaska Range, before plunging back down south and leaving Fairbanks in frigid night. Temperatures regularly plummet to −40°C. It's so cold that the air behaves differently: distant sounds become eerily close, and the smoke from home fires lies horizontally across the rooftops.

Or at least it used to be that cold. In recent winters temperatures in Fairbanks have reached −30°C for only a couple of days, Dale Curtis told me, whilst in previous decades they had remained at −40°C for months at a time. And it's not just Fairbanks: similar stories come from all over the state. In Barrow, hundreds of kilometres above the Arctic Circle on Alaska's frozen north coast, there were thunderstorms for the first time in memory a couple of years back. Local people had never seen anything like it: some Native American elders thought that the loud bangs of thunder were bombs going off.

The reason is simple: Alaska is baking. Temperatures in the state – as in much of the Arctic – are rising ten times faster than in the rest of the world. And the effects are so dramatic that entire ecosystems are beginning to unravel, as are the lifestyles of the people – many of them Native Americans – who depend on them. In many ways Alaska is the canary in the coal mine, showing the rest of the world what lies ahead as global warming accelerates.

*　　*　　*

The man who has done more than perhaps any other to highlight this is a quietly spoken scientist based at the University of Alaska in Fairbanks, Professor Gunter Weller.

I met him on a warm May morning, and the season's first mosquitoes were descending from the trees as we walked through patches of thawing snow behind Weller's Center for Global Change and Arctic Systems Research. The Center looked thoroughly modern, with shiny plate-glass windows and a big car park full of big sports utility vehicles. A hundred metres away in the forest, a giant dish was turning gradually around, focusing its data stream on some unseen satellite far above.

'In Alaska we're seeing great changes in climate,' Professor Weller was saying in his soft German accent. 'There's no doubt about it. This year we had extreme high temperatures, and in fact it's been the second warmest year on record.'

When he first arrived in the state, Weller continued, the weather had been quite different – sometimes reaching minus fifty centigrade. 'In fact, I remember my first New Year's Eve here in 1968. I was invited next door to a party and I put a shot of very good scotch in an ice-cube tray outside, and it was frozen within half an hour. You wouldn't see that now, no way.'

Alaskan wintertime temperatures have shot up by an average of six degrees centigrade, Weller told me. 'This is an absolutely enormous signal,' he emphasised, 'bigger than anything the computer models have predicted.' Summer temperatures were rising too: Fairbanks now regularly sees summertime highs of twenty-five degrees.

One of the best temperature records of all comes not from scientists but from gamblers. Each year the people of Nenana, a small town southwest of Fairbanks, place bets on the exact minute when the ice on the river will begin to break up for the spring thaw. The contest began when Alaska Railroad engineers put down an $800 wager in 1917; by 2000 the jackpot had grown to $335,000, and thousands of people across the state compete. The high financial stakes ensure constant vigilance by the locals, so the record is considered as reliable as the best scientific data – and it shows that the first day of spring has advanced by over a week since the 1920s.[2]

So was this global warming? I asked Professor Weller. Or perhaps something else?

His answer was unequivocal. 'I think it's clearly under-stood and clearly accepted by the scientific community that this is in part due to the human-induced global green-house effect.' This greenhouse effect, he explained, was amplified at high latitudes by a positive feedback: once snow and ice begin to melt, the reflectivity of the Earth's surface decreases, allowing more of the sun's heat to be absorbed. This in turn melts more ice and snow, further reducing the planet's albedo, allowing still more warming, and so on.

In Fairbanks the rising temperatures were having a dra-matic impact. Much of the area is underlain by permafrost – permanently frozen ground – which now, for the first time in thousands of years, is beginning to thaw. As a result, houses are sagging, roads collapsing, and entire buildings being swallowed up by holes in the ground.

Professor Weller gave us a lift to one of the worst-affected neighbourhoods in the city – the aptly named Madcap Lane, where most of the wooden one-storey properties were sagging and distorted in several directions. On the right-hand side one house was tilting sideways, the guttering at one end about a foot further from the roof than at the other. The wonky front steps barely fitted into the porch. I climbed them carefully and knocked on the door.

'I work nights, and I've just gone to bed,' complained the woman who opened it, glaring at me through sleepy eyes, hair standing up in all directions.

'It's our house, it's not a museum,' added her teen-age daughter irritably from behind the half-closed door. I backed off, and let Franny and Karen take over. Within a few minutes they were deep in conversation with the woman, Vicki Heiker, who soon invited us inside.

Vicki's daughter Jessica was also smiling now. 'Here, look at this.' She placed a pencil at one end of the kitchen table. It quickly rolled off the other end onto the floor.

'Can you do it again with me filming?' asked Franny, and Jessica happily obliged. Her mother watched them and laughed. 'When you spill something it's like you don't have much of a chance. You've got to clean it up fast otherwise it'll get away from you.'

'Do you get used to it?' I asked.

'Well, it helps build up your calf muscles since you're always walking uphill.' In the sitting room a large dog was jumping around excitedly, shedding dark hairs all over the furniture. Vicki bent down next to a stack of glass shelves in the corner, and pointed to an inch-thick triangle of wood under one of its legs.

'That's a shim. It's elevating that leg to make the whole thing level. Otherwise it would just fall over.'

'Sometimes my bedroom door won't shut all the way either,' Jessica chimed in. 'Here,' she went on, changing the subject, 'do you want to see my pet ferret?'

'I don't think anything here is entirely level,' continued Vicki, ignoring the struggling animal, which broke free from her daughter's grasp and bolted behind the sofa. 'You've seen the front porch? Well, my son had to come round last week because it fell off.'

I wandered back into the kitchen, where Karen was taking pictures of an impressive-looking crack which snaked out from above the kitchen window. The window was itself askew, and looking through it I could see the house across the street also tilting – in the opposite direction. The whole place was like a badly built Toyland.

'It's a trip, isn't it?' smiled Vicki, coming up beside me. 'You look out of here and it makes you feel dizzy.'

But Vicki and Jessica's house was far from being the worst example. On the way back to our guesthouse the taxi driver took us on a detour up a very narrow track lined with dirty snowbanks. At the end was a log cabin which looked as if it had been pitched into a gigantic hole: it was tilted at an angle that was at least 20 degrees.

'I stuck it out until two years ago,' said the elderly woman who had been forced out into a different house nearby. 'The water and electricity still work, you know. And when my husband first built it back in 1957 the ground was completely flat.' She shook her head in disbelief.

Roads all around Fairbanks are affected by thawing permafrost: driving over the gentle undulations is like being at sea in a gentle swell. In some places the damage is more dramatic – crash barriers have bent into weird contortions, and wide cracks fracture the dark tarmac. Permafrost damage costs now total $35 million every year, money mostly spent on repairs to affected roads.[3]

Forests are affected too. Some areas of once-flat land look like bomb-sites, pockmarked with craters – sometimes several feet deep – where permafrost ice underneath them has melted and drained away. These uneven landscapes cause 'drunken forests', a phenomenon that has been reported right across Alaska. I saw plenty of evidence of this around Fairbanks: in one spot a long gash had been torn through the tall spruce trees, leaving them toppling over towards each other. Most of them were dying, some already lifeless, their brittle branches snapping off as I pushed through to take a closer look.

Permafrost degradation is one of the clearest signals that something unprecedented is happening in the far north. In Siberian cities hundreds of tall buildings have begun to subside and crack.[4] Whole ecosystems are disappearing too: the Boreal forests which have grown throughout the region since the end of the last Ice Age are now collapsing into marshy bogs as the ground underneath thaws out. In

Alaska, spruce and birch forests are being replaced by wetland – in some areas a quarter of the forest has disappeared in the last forty years.[5]

Whole sections of coastline are breaking off and falling into the sea, as the ice which has kept cliffs solid for centuries begins to melt. In both Canada and Siberia losses of up to forty metres a year have been observed, and in Alaska over half a kilometre has eroded from some stretches of coastline over the past few decades.[6] This may not matter too much when nobody lives there – but many of these coastal areas have been inhabited by indigenous peoples for centuries.

And in Shishmaref, on the west coast of Alaska, the Native Americans who have lived on the same site for decades now live in daily terror of the sea.

SHISHMAREF

It was impossible to tell that Shishmaref was even on the coast. Although the village is actually squeezed onto a long, narrow barrier island, all I could see as our small aircraft looped over the area was a grey airstrip and about a hundred houses in the middle of an immense white plain. A layer of heavy grey cloud hung over the entire area, meeting the horizon in an apparently infinite expanse of nothingness. On the ground it was chilly to say the least: the temperature hovered around minus fifteen Celsius, and a few snow grains blew in the biting northwesterly wind.

Shishmaref is about as far west as you can get on the entire North American landmass. The tip of the Seward

Peninsula, on which the village sits, is barely a hundred kilometres from the eastern edge of Russian Siberia. The International Date Line runs through the middle of the freezing Chukchi Sea which separates the two coastlines, meaning that the same morning sun rises a whole day later on the Alaskan side.

The two landmasses are so near that their peoples are closely related too. Almost all Shishmaref's residents are Inupiat Eskimos, who share a close language and ancestry with their Siberian Eskimo relatives. (Unlike in Canada and Greenland, the name 'Inuit' never caught on in Alaska, and the terms 'Eskimo' and 'Indian' are still universally used to describe the two culturally distinct Native Alaskan first peoples, both by themselves and by Alaskans of non-Native descent.)

Indeed Alaskan Eskimo hunters, cut off from home by open-water leads appearing behind them in the sea ice, would sometimes accidentally spend entire summers in Siberia. A lost hunter's family would never give up hope until the following winter, when men who had survived would return back over the newly frozen ice.

You still have to be careful out on the ice, Robert Iyatunguk, Shishmaref's 'erosion co-ordinator', told me as he showed me round the village. Anyone who falls through into the water has only minutes to strip completely and change into dry clothes before they freeze to death. And sometimes the ice does strange things: 'I was once out there on an ice floe with some friends and got this weird feeling of danger,' he recalled. 'We all cleared off and immediately the whole floe started to turn over.' There is safety in numbers – no one goes out alone, and a group of hunters will always share the harvest equally.

Until comparatively recently Shishmaref's entire food and clothing supply came from the surrounding environment: polar bears, seals, fish, walrus and caribou. Though dog sleds and bone arrows have now been exchanged for snowmachines and guns, and Eskimo kayaks replaced by wooden or fibreglass boats, 'subsistence' living remains a crucial part of people's culture and livelihood. Bits of hunted animal – a frozen caribou leg or part of a seal – were propped up around almost every doorstep, and polar bear skins and dried fish hung on racks behind the houses. Not far from our temporary lodgings at the Lutheran pastor's residence, a severed musk ox head stared at the grey sky through clouded, lifeless eyes.

A few decades ago people lived in 'sod houses', turf-roofed dwellings dug out of the ground, dark and dingy but very well insulated from the winter cold. A few lumps further up the shore are all that remains of them – today everyone lives in wooden or prefabricated modern homes, scattered in rows all around the island.

Nine houses had to be moved during the last big storm, Robert Iyatunguk told me. As ninety-miles-per-hour winds whipped around them, and whole sections of thawed cliff tumbled into the raging sea, the whole community had mobilised to save the dwellings which were closest to the edge. It was dangerous work: not only could a house collapse on the people working under it to jack it up, but the ground itself could give way suddenly beneath them.

'We lost fifty feet of ground in one night with that storm. We're in panic mode now because of how much ground we're losing.'

We crunched down a shallow slope where sandbags were protruding through the snow: the remnants of Shishmaref's last battle with the sea. All the sea walls had failed, he went on. The water just undercut or washed over them. It seemed like nothing could prevent this loose, gravelly ground from eroding away.

Now the talk was of relocation – something that would have to be agreed by all 600 residents through a community ballot. (It was – over a year after I'd left – in July 2002.) It would cost $50 million, and there was no sign of the state authorities coming up with the cash. But the worst case scenario was no longer that of having to move the village, he said, but that of another big storm whilst they were still living in the danger zone.

Time is running out, Robert emphasised. 'The wind is getting stronger, the water is getting higher, and it's notice-able to everybody in town. It just kind of scares you inside your body and makes you wonder exactly when the big one is going to hit.' And this 'big storm' throws a perpetual shadow over the community the longer it stays put: people cast anxious glances over at the horizon, and when a strong wind gets up, those closest to the shore often decamp to sleep at relatives' houses.

There was an emergency evacuation plan of sorts – something partly within Robert's responsibility that has given him many sleepless nights. In a few hours a C-130 aircraft could arrive, and evacuate all of Shishmaref's residents within four return journeys. But could it operate during a storm? And what if the runway began to collapse? 'If our airport runway gets flooded out and eaten away, there goes our evacuation by plane,' Robert admitted. 'Then we'd have to go to the next highest point in town, which would either be the church or the school.'

We stood together under the crumbling cliffs. Robert scuffed the base of it with his boot, and icy sand showered down. Up above us an abandoned house hung precariously over the edge, at least a third of its foundation protruding into thin air. The house next door had toppled over and been reduced to matchwood by the waves.

'There's one major storm that we never had,' Robert Iyatunguk concluded quietly. 'I'd hate to be here when it hits, but my kids are here, and I'm going to stay here with my kids and my wife's family and their brothers and sisters for as long as it takes.'

I spent that evening with Clifford Weyiouanna, a fifty-eight-year-old Shishmaref elder, who sat polishing his gun as we spoke. Several snowmachines were parked outside his house, one with a sled on which were stacked three large blocks of ice – clear as glass, and cut from a coastal river to serve as drinking water. Children were playing on the snowdrifts – some piled as high as the houses themselves – and Clifford's grandchildren ran in and out of the house, banging the door behind them.

'It's no good getting old without kids around,' Clifford chuckled indulgently as one of them whizzed past. He brought out an 'Eskimo shotgun' (a bone harpoon which had been used to hunt ducks) and his most prized possession, an intricately-beaded woven belt, so ancient that no one remembered who had made it.

Having lived in the community all his life (bar four years at high school and two in the military 'hellhole' of Fort Benning, Georgia, and then Saigon, which he didn't like to talk about), Clifford Weyiouanna was an authority on the local environment. It was true, he told me, that the permafrost underlying the village was melting, and this was speeding up the erosion. But another factor was just as important – the gradual disappearance of the sea ice.

The sea ice used to lock up the shore for six months of every year, he explained, and so for half the year the eroding power of the waves was banished. Storms could rage all they wanted, but the sandy cliffs would stand. Now that had begun to change.

'The currents have changed, the ice conditions have changed, and the freeze-up of the Chukchi Sea out here has really changed too. We used to freeze up in the last part of October. This year we didn't freeze up until Christmas time.'

'So, how different is it when you're actually out on the ice?'

'It's not as stable. We used to get icebergs from the north many years ago – turquoise blue icebergs – not any more, it's all young ice now. Thin stuff, only about a foot thick.

Right now, the ice on that ocean out there should be, under normal conditions, four foot thick.'

And the animal behaviour was changing too. 'I think they're migrating a lot earlier than they used to because of the warming of the ocean. They migrate north in the spring to stay in the cooler waters. That's the polar bears, the walrus, the spotted seal, the bearded seal, the belugas and the bowhead whales.' He leaned forward to emphasise the point: 'Last summer we covered thousands of miles by boat trying to get walrus – there was nothing, except for one boat which found one walrus.'

And then there were the strange new fish. 'I used to have one in my shed. I was going to give it to a biologist to take a look because it's not a local fish. The warming of the temperature is bringing some uncommon fish species into the ocean.'

We talked long after midnight. Outside the sun was only just setting, and the kids were as noisy and energetic as ever. No one bothered to order them around: traditional teaching methods are subtle, and Eskimo children are expected to find things out for themselves.

Shishmaref would go on, both Clifford and Robert assured me. If not here, then someplace else further up the coast. But whatever happened, the community would stay together. People here looked after each other – just as the first seal of the hunting season would always be given to an elder. It was the traditional way.

HUSLIA

All over the Alaskan interior people in remote villages are reporting sudden changes, all related to the state's warming climate: weird animal behaviour, unexpected weather, changing landscape and dying forests. Around Huslia, a small Athabaskan Indian village three hundred kilometres west of Fairbanks, entire lakes have disappeared.

These disappearing lakes sounded a bit too dramatic, and I wasn't sure I believed it – until I visited the village and saw it happening for myself.

The plane was only an eight-seater, and I was directly behind the pilot. The dials spun as he heaved back the joystick, the small craft gaining speed and then bouncing into the air from a side runway at Fairbanks Airport. Soon we were flying over thick forests, which encircled huge ox-bow lakes formed by old river courses. As we cruised at only 900 metres, thin ice clouds scattered the bright sunlight into an ever-present rainbow on the left, whilst on the right, small mountains rose above the treeline, looking almost impossibly smooth under their thick coating of snow.

Huslia was over two hours away, first visible as just a little grey airstrip and a few dozen cabins as we glided in over the forests. As a Native village, Huslia has its own Tribal Council, and one of the officials was waiting to meet us. We loaded our bags onto a sled and rode down into the village on the back of her snowmachine, drawing to a halt outside a log cabin with a large freezer outside the front door and lots of toys scattered around it in the snow.

Cesa Sam appeared at the door. She was dressed only in a T-shirt and shorts, despite the cold weather. Inside, I could see why – the house was boiling, and we all rapidly stripped off our coats and gloves. Cesa was in her early thirties, large and cheerful, and continually pestered by several hyperactive children.

'Oh, a lot of people can tell you about the weather,' she said, when we were all drinking hot chocolate around the table. 'There's only been one cold winter since '94 here. It's so much warmer, and that's a big change.'

I went for a walk around the village that evening. It was bordered on one side by a wide frozen river, with steep banks leading down to the ice edge. Lots of snowmachine tracks led along the river, which was clearly the equivalent of a main road in winter. But the spring break-up was just beginning, and dark patches in the snow indicated where water was seeping through from the thawing ice underneath.

Everyone from the very young to the very old seemed to get about by snowmachine, and the saw-like buzzing of motors made a constant background noise. There were fewer kids around than normal: I later found out to my surprise that most of the school seniors had gone on a trip to Mexico. Demolishing my assumptions about Indian villages, modernity was everywhere – televisions flickered inside most of the houses, and on a makeshift basketball court two wiry teenagers were sliding about on the ice, taking turns shooting the ball. Like other kids I'd seen elsewhere in the United States, they wore baggy jeans and sneakers, and moved with a disinterested, thoroughly urban cool.

As in Shishmaref, subsistence food is still vital. At a 'potlatch' communal meal later in the evening, hunks of caribou shared space with jelly, ice cream and crisps on paper plates. The elders played bingo several nights a week, sitting attentively at classroom desks in the community centre. The Huslia village store was packed with dried soups, big plastic bottles of coke, biscuits and even some fresh vegetables like onions and carrots, a new shipment of which had come in on our plane. But in the summer the whole community moved out to 'fish camps' to catch salmon, and the traditional diet again predominated. Cesa's own house, where we stayed in an upstairs boxroom, doubled as the village video store, and was well patronised by residents seeking repeat viewings of Eddie Murphy films and *Titanic*.

Although village life looked relaxed enough, the relationship between modernity and traditional lifestyles is never easy – in Huslia as in other Native villages across Alaska and the United States generally. The Koyukon Indian language – part of the Athabaskan language group, that includes the Apache and Navajo as far south as Arizona and California – is dying. Old people still speak it to each other, but the middle generation were beaten by their white schoolmasters if caught speaking it, and everyone now speaks English at home.

Alcoholism is a huge problem, even in 'dry' villages like Huslia, and several recent teenage suicides have shaken the community's confidence to the core. No summary can

explain the social crisis underlying this kind of tragic behaviour, but loss of culture is surely a central problem, contributing as it does to the breakdown of community values
and roles, alienation, loneliness and poor self-esteem.[7]

In a way, these wider cultural changes ran parallel
to changes in the surrounding environment. In the past
people derived meaning from the regular progression of the
seasons – from the migration of the caribou to the first
appearance of the salmon in early summer. These rhythms,
and the subsistence lifestyle generally, explained the world
and made the people feel part of it.

But now the salmon sometimes failed to appear on
time, and the previous year all the berries died before they
got ripe. Hungry bears were ranging closer to the village.
Willow trees were springing up where there used to be
standing water, and most of the beavers had disappeared.
The world was unravelling, and even the most stoic and
experienced elders were at a loss to explain what it meant.

And underlying everything was the rising temperature.

'Right now we hardly see forty below all winter.' I was talking to Wilson Sam, Cesa's father, the following morning
in his kitchen. 'I think we maybe saw one day of it, but the
rest was like twenty-five, thirty below. And that's all winter,
that's a big change.' Wilson and his wife Eleanor were
plucking geese, plunging the dead birds into boiling water
to loosen the feathers then tearing off great handfuls and
piling them up on the kitchen table. Wilson had shot over
a dozen the day before.

'My parents used to have really warm gear,' said Eleanor. 'I remember my late father, he had long caribou legging boots about this high.' She put down her half-plucked goose to indicate. 'All us children, we had fur coats too – real fur coats. My mother had a rabbitskin parka. Now the weather's really changed, and people don't use that kind of fur clothing so much any more.'

'Now if it gets to forty below people say it's cold,' added Wilson. 'But in them days it was colder. And it lasted for days sometimes. Worse than, what, fifty, sixty below. You know, real cold.'

Eleanor looked up, as if she had just remembered something. 'My grandpa, he said in our Athabaskan language before he died, when he was in his eighties. He said the cold weather is going to get old. Because it's getting warmer in Alaska, you know? The cold weather's going outside.'

That afternoon I was riding Cesa's snowmachine down a steep slope, trying to keep up with Harold 'Farmer' Vent, a Huslia old timer and councillor. Farmer looked like he'd seen a good few Alaskan winters: his lined face tucked under a pine marten skin cap, the buff-coloured tail hanging down the back, he looked every inch a skilled trapper. Always about fifty metres ahead, he kept disappearing around stands of forest and behind clumps of bushes, and I was worried about losing him. I had no idea which way led back to the village, and the landscape of forests, snow-covered depressions and riverbanks all looked identical.

Then, abruptly, Farmer drew to a halt. 'This is it,' he announced.

We were in a large bowl-shaped area, a kilometre or so across. Much of the snow had melted, leaving dusty grass and a tangled mat of dried-up pondweed. It was only then that I realised, with a jolt, that this had once been a lake.

'The water's just draining out,' Farmer said. 'I don't know where it's going. We used to paddle down here in canoes during the summer to get to my mom's fish camp. We got to carry the canoes now.'

The area around Huslia used to be covered with lakes. 'Every spring they still fill up with water, but then it just drains out – all the way to the bottom. All these lakes are drying up now, they're just grass.'

He climbed back onto his snowmachine, and I followed him for a couple of kilometres more – up a steep bank and then down the other side before he stopped again. The scene was the same, though this time a line of birch trees surrounded the dusty hollow, indicating what had once been a lakeshore.

'It's all over the area,' Farmer told me. 'I trap way up towards Hog River, and all those lakes are drying out too.'

I asked what difference it made to the animals.

He shook his head sadly. 'Ducks, beaver, muskrat ... We used to shoot muskrat off this hill right here, but everything is drying out, so we can't get nothing. With beaver it's the same thing.' He pointed to the edge of the bank. 'There used to be a beaver house right over here. They're all moving someplace, I don't know where.'

We stood in silence, as Farmer stared at the ground. 'It's just – what do you call it? . . . Pitiful really. Even the geese and stuff, they've started disappearing now. Every year it's getting harder and harder to live up here.'

Polar warming

Evidence of dramatic climate change is piling in from right around the Earth's polar regions. Greenland's ice sheet is thawing so fast that meltwater is running off at a rate equivalent to the annual flow of the Nile.[8,9] Throughout the Northern Hemisphere winter snowcover has declined by a tenth since 1979,[10] whilst rivers and lakes are freezing a week later and thawing a week earlier than a century ago.[11] Alaskan mountain glaciers are losing ice at a rate fast enough to have a measurable impact on global sea levels,[12] whilst other glaciers and snowfields are disappearing throughout the region.[13]

This warming is mirrored in the Southern Hemisphere, where the Antarctic Peninsula is warming at a rate similar to that in Alaska. (The Antarctic continental interior, which is surrounded by cold circumpolar winds and sea currents which isolate it from wider global temperature changes, has warmed much more slowly, if at all.) As a result, snowcover and glaciers on the Peninsula are shrinking, ice-dependent Adelie penguin populations shrinking and new plants are beginning to colonise the landscape.[14]

About 10,000 square kilometres of ice shelf have been lost from both sides of the Antarctic Peninsula, culminating in March 2002 with the spectacular collapse of the Larsen B ice shelf, an event which made headlines around the world. Before its sudden demise, Larsen B was a floating wedge of ice 200 metres thick and larger than the entire country of Luxembourg. 'The speed of it is staggering,' said a British Antarctic Survey glaciologist at the time, as his ship navigated through the armada of new icebergs. 'Hard to believe that 500 billion tonnes of ice sheet has disintegrated in less than a month.'[15]

Much of what I had heard from Alaskan residents is backed up by hard scientific evidence. As Clifford Weyiouanna told me, sea ice is thinning rapidly. This observation is confirmed by submarine cruises under the Arctic ice, which reveal a thinning trend of over 40% over the last thirty years.[16] The total area of Arctic sea ice is also diminishing rapidly: satellite data shows an area one and a half times the size of Wales is lost every year.[17] In September 1998 ice cover in the Beaufort and Chukchi seas around Alaska (Shishmaref is on the Chukchi Sea) reached a record low, a quarter less than the previous minimum extent over half a century of observations.[18]

As Clifford and other Shishmaref Eskimo residents told me, this reduction in sea ice is bad news for marine life, much of which congregates around the edges of the ice, where multitudes of plankton and fish form a food bonanza. Walruses, for example, need sea ice thick enough to hold their weight but in shallow enough water to allow them to dive and feed on the sea bottom. Similarly, ringed seals depend on sea ice as a habitat for pupping, moulting, foraging and resting. The same is true for many Arctic species: the health of populations of walruses, ringed and bearded seals, polar bears, belugas and bowhead whales are all strongly correlated with sea ice cover.

The reported changes on land are also well supported by scientific research. Satellite pictures of the Alaskan interior confirm that lakes and ponds have been drying up during the last decade.[19] The IPCC suggests that this phenomenon is linked with melting permafrost: frozen ground forms an impermeable layer, but once it thaws, surface lakes can drain away.[20] As I had discovered in Fairbanks, widespread permafrost melting is well underway across large areas of Alaskan territory, affecting not just buildings and roads, but also wild forests.

Huge areas of woodland have also been destroyed by another side-effect of warming – spruce-bark beetle infestations, which have killed 2.3 million acres of trees since 1992 across a broad swathe of southern Alaska. The devastated area reaches right to Anchorage itself, and visitors flying into the city's airport cross islands covered with the bristling, white skeletons of dead trees which are easily visible through the plane windows. It's the worst insect outbreak ever to hit North American forests, and is directly related to higher temperatures: in colder winters, the beetle eggs had been killed off and the population had been unable to explode.[21]

On the southern Kenai Peninsula, an area famed for its undisturbed natural forests, the beetles attacked like a plague of locusts. One local wildlife specialist compared it to 'an Alfred Hitchcock movie'. As he told *Alaska Magazine*:

'They would be in your hair and eyes, you'd have to brush them off. I've heard people saying they could see them in clouds, miles off, coming down the valley.'[22] And the only thing that stopped the plague was when there were no more trees left alive to attack.

PRUDHOE BAY

The spruce-beetle outbreak, the destruction of forests, buildings and coastlines by melting permafrost, and the disappearance of sea ice are a disaster for Alaska's people, wildlife and natural heritage. So who is to blame? Partially – and here lies the irony – the chain of causation leads straight back to Alaska itself. Oil extraction has dominated Alaskan industry for over twenty years, and this oil has been contributing directly to rising temperatures through the greenhouse gases released into the atmosphere during its combustion.

You'd be hard pressed, however, to find anyone in Alaska prepared to admit this. People know which side their bread is buttered on, and with 80% of state revenues coming from royalties paid by drilling companies,[23] and many of the highest-paying and most reliable jobs based on extraction and oilfield services, no one wants to rock the boat.

Oil money has poured into the coffers of state politicians, with both Democrats and Republicans competing to offer the industry tax breaks and other incentives.[24] And ordinary Alaskans benefit too – every year every state citizen, from the oldest granddad to the youngest baby, gets a payout from the Alaska Permanent Fund, a state fund now totalling more than $20 billion, collected from decades of oil company royalties. In 2002 the APF dividend cheque came to $1500, free money for everyone, and a convincing reminder of the rewards paid by Big Oil.[25]

Many articulate environmentalists have found a place in Alaska, but they are marginalised and vilified by the political establishment, and the Prudhoe Bay oilfield is a no-go area for anyone identifying themselves as a 'green'. Greenpeace ran a long battle against a new BP offshore facility in the Prudhoe Bay area in the late 1990s, but was practically run out of town by a coalition of local Eskimos and oil drillers.

More recently the debate about whether the Arctic National Wildlife Refuge should be opened up for oil extraction has polarised the situation still further. Although polls show that most Americans want the Refuge protected, Alaskan politicians almost unanimously demand it be opened up. Concerns have been raised by the Gwich'in Indian tribe and others that oil drilling would destroy the calving grounds of the Porcupine caribou herd, but this claim is strenuously denied by politicians and oil companies alike.

The coastal plain of the Refuge, under which somewhere between two and ten billion barrels of oil are thought to lie, has been called 'America's Serengeti'. According to the US Fish and Wildlife Service, the agency responsible for managing the Refuge, it is vital not just for the caribou, but for golden eagles, snow geese, polar bears, lynx, muskoxen, arctic foxes, wolverines, grizzlies and countless other species.[26]

None of this dewy-eyed, liberal concern cuts any ice with Alaskan businessmen or politicians. Indeed, a well-funded lobbying group in Anchorage called Arctic Power exists expressly to campaign for opening up the Refuge, and receives donations not just from the oil industry but directly from taxpayers via the state budget (the 2001 state appropriation totalled $1.8 million[27]).

Knowing that this was among the best places to hear an oil driller's view of the situation, I visited Arctic Power's offices.

The director Cam Toohey is as Alaskan as they come: born and raised in the fishing community of Homer, he is also a keen 'musher', and raced sled dogs for fifteen years. A framed photo of his wife and two blond boys next to a swimming pool was propped up on the desk, and his office walls were decorated by Eskimo face masks and posters of colourful Arctic sunsets. He was personable and chatty, and wore a perpetual smile – until I asked him about global warming.

Then he looked uneasy. 'Well, you have to understand that 10,000 years ago we were in an ice age.' I should therefore realise, he informed me, that climate changes were natural and happened all the time.

'Yes, but do you accept that the human-enhanced greenhouse effect is currently underway and having an impact in Alaska?'

'Well, I think the jury's still out about how much of a contribution the public has made to the greenhouse effect in their consumption of fossil fuels. No one has determined that we can stop consuming fossil fuels today and still have a healthy environment and a healthy economy.'

I was mystified as to how consuming fossil fuels was necessary to maintain a healthy environment, but decided to let it pass.

Instead, I tried a more emotive tactic. 'The latest predictions coming from the scientists at the University of Alaska in Fairbanks suggest catastrophic warming for Alaska over the next century. This is where your kids are going to grow up. Doesn't that worry you?'

He repeated once again that temperatures had been changing for thousands of years. 'Obviously if science is able to determine that this greenhouse effect is being caused by human contribution then I'm sure the nation and the world will do things to try and address that. But it is not a clear-cut scientific situation right now.'

I had given up pretending to be neutral. 'Yes it is,' I insisted, referring to the three landmark reports published by the Intergovernmental Panel on Climate Change, the latest in 2001.

Toohey hadn't read any of them. 'But I can tell you that in the US there is not a consensus on what the solution is to climate change – nor what the causes are,' he maintained. And, in the meantime, he concluded, it was better to drill for oil on home territory than to depend on unstable dictatorships in the Middle East. So the Arctic National Wildlife Refuge had to be prised open.

I had one last question. 'After that, will you be happy,

or will you be campaigning to open up somewhere else?'

'Well, it all depends,' he smirked. 'We have other regions in the state that hopefully can be utilised for natural gas production. As long as we have the resources – and there's a demand in the country for production – we should be able to develop it.'

I left feeling unsatisfied. Cam Toohey was clearly no dummy – his answers were articulate and relatively well-informed. Yet I had been met with a wall of denial. There was no sign of doubt, nor any suggestion that precaution might be a good policy given the potential magnitude of climate change. There was no alternative, he had said. Economic development must march forward, whatever the weather.

Since I conducted that interview, Cam Toohey has been appointed by the Bush administration's Interior Secretary Gale Norton as her Special Assistant for Alaska. Toohey, the *Alaska Oil & Gas Reporter* noted ominously, would 'assist Norton with the management of 270 million acres of federal lands that fall under the jurisdiction of the Department of the Interior'.[28] Clearly his appointment reveals something about the seamless connection between the oil industry and the current US administration. Indeed, Norton herself has a distinctly un-ecological track record, having spent much of her earlier (legal) career fighting environmental regulations and promoting the interests of polluting corporations.

Not everyone was pleased by the decision – one Congressman said that Norton's new Assistant was an 'ethical oil spill'.[29] Prescient words, it seems: during the first ninety days of Toohey's new role, he was engaged in allowing increased numbers of tourist cruise ships back into the fragile Glacier Bay, despite a court decision restricting their numbers on environmental grounds; weakening Clinton-era mining restrictions; promoting new oil-drilling leases in offshore Alaskan waters; and, of course, working to achieve his lifetime ambition of opening up the Arctic National Wildlife Refuge.[30]

Clearly Alaska's public lands are in extremely attentive hands.

Having had my first lesson in the oil industry mindset, I felt it was time to visit the heart of the beast – Prudhoe Bay itself, the largest oilfield ever discovered in the Western Hemisphere. Located on the northern Arctic Ocean coast of Alaska, Prudhoe Bay is only one of a whole complex of different oilfields, all stuck out in the freezing flat tundra of the North Slope. With the sea coast ice-bound for most of the year, the oil is transported 1200 kilometres south to the warm-water port of Valdez via the Trans-Alaska Pipeline.

One of the biggest industrial developments on Earth, the scale of the North Slope oil development is only really apparent from the air: as my plane came in to land, literally scores of drilling pads – each a set of little box-like buildings arranged around a central rig – stretched as far as the eye could see across the monotonous white plain.

The whole thing takes up about 650 square kilometres (250 square miles), and stretches east–west along the shore of the Arctic Ocean for nearly two hundred kilometres. Each well (there are up to forty wells per drilling pad) sucks oil from several thousand metres down and over an area of 80 acres or more. Every drop of this oil eventually ends up being burned in the cars, trucks and aeroplanes that keep America's economy turning.

Most of the Prudhoe Bay area is off-limits to the public, officially for security reasons. It was bitterly cold when I arrived at the small town of Deadhorse, and I hurried through the powder snow over to the Prudhoe Bay Hotel, the only accommodation open to visitors.

The hotel was basic, with long corridors of rooms sharing the same washing and toilet facilities throughout the lowslung two-storey building. Tough-looking workmen, all wearing grubby jeans and baseball caps, stomped up and down the hallways, fresh from the outlying drilling areas – many on their way back down south for breaks with their families. Food was included in the room price, but a sign advised hotel residents that breakfast was only served between 4.30 and 7.30 a.m. Another handwritten sign on the door warned that bears were in the area.

BP, the largest operator in the Prudhoe Bay area, had spurned my request for a guided tour, but I was able to get in anyway thanks to the deputy manager of the hotel, who ran his own minibus tours. As we bumped along the gravel roads, having shown our passports to a guard at the checkpoint, he explained the Prudhoe Bay lifestyle. Although about 1500 people worked in the oilfields, he said, there were only twenty-five permanent residents – most in

Deadhorse itself. Everyone else worked shifts: two weeks on, two weeks off. 'When you're on shift you work twelve hours a day, seven days a week. When you're off, you fly home.' No alcohol was permitted in the entire place – not even in the hotel.

We passed a large yellow building belonging to Halliburton, the controversial oilfield services company whose former CEO is the current US vice-president Dick Cheney. 'Dick was here just a few weeks back,' recalled my guide nonchalantly, before slowing down to indicate the Central Gas Facility, a massive red complex the size of three football stadiums.

Then we were at the shore of the Arctic Ocean itself: a small slope before an impossibly large, white expanse. 'The North Pole is over there,' he indicated, before pointing out BP's Northstar oil platform twelve kilometres offshore. I tramped through the powder snow onto the ice itself, savouring the moment. Small snow flurries fell from an almost-blue sky, and large cracks were visible where the sea had pushed ice up against the shore.

My guide was muttering darkly about the Greenpeace campaign which had focused on Northstar a few years previously. They were 'very rude people', he recalled, who had been airlifted in because none of the native Eskimos would help them reach the site overland. No one would give them space in the hotels either, he said with a chuckle. 'They caused all sorts of problems. And you know what?' he concluded triumphantly. 'They left the place in such a mess that the oil industry had to clean up after them!'

On the way back we detoured via Pump Station One –
the very beginning of the Trans-Alaska Pipeline. There was
a section cut out of the outer pipe-casing so people could
put their hands in: the inner pipe was warm, like blood, and
vibrated slightly as the black liquid surged away southward.
This was the same oil, incidentally, which poured out of the
Exxon Valdez tanker in March 1989 in one of the world's
worst oil spills.

Back in the hotel I was stopped in the corridor by Max, a
half-Eskimo guy, who regarded me suspiciously. 'You from
Greenpeace or something?'

No, I was a journalist, I told him.

He assumed I was covering the Arctic Refuge story –
the western border of the Refuge is not far from Prudhoe
Bay. 'I'd say that the oil companies have supported a lot of
people for a long time,' he began. 'In the 1970s we were
having a hard time. I remember having to put our pennies
together to even buy a can of soda.'

I asked him about global warming.

'I think it's all hype,' he snorted. 'We had snow in the
first week of May in Fairbanks.'

An older grey-haired man, a policeman from Barrow,
joined in. 'Global warming don't come from here, it comes
from Chicago, New York, where all the emissions are.' He
jabbed his finger forcefully. 'The whole world depends on
oil, so why are we always the bad guys?'

'You came over on a jet, right?' demanded Max. 'Or was
it solar-powered or electric-powered?' He fixed me with

a mocking grin. 'You sure you're not from Greenpeace?'

All these mentions of 'Greenpeace' were beginning to draw a crowd. I backed off, using the excuse of fetching my press card to show Max.

The policeman followed me. 'The air here's clean,' he persisted. 'You can't even burn waste without monitoring emissions.' He turned to a grizzled man in a Chevron baseball cap: 'What do you think about global warming?'

The man had just come in from outside, and looked cold. 'I'm ready for it!'

Like most outsiders, I have long been conditioned to think that indigenous people usually fight against the oil industry, so finding out that the North Slope Eskimo communities were some of the industry's strongest supporters initially came as a shock.

As the North Slope Borough Mayor George Ahmaogak puts it in a glossy brochure I was given by Cam Toohey's secretary:

> As Mayor, I can state unequivocally that the people of the North Slope Borough enthusiastically support the presence of the oil industry in our land. North Slope oil has already provided immense benefits to our people and to our country. Well-meaning Americans crusading against Coastal Plain development would deny us our only opportunity for jobs – jobs providing a comfortable standard of living for the first time in our history.[31]

In my last stop before leaving Alaska, I was particularly interested to hear how oil development could be squared with the widespread Native American view of themselves as custodians of the land, and whether anyone was noticing the impacts of global warming or knew any details about its cause. It was time to visit one of the closest Native settlements to Prudhoe Bay, and the only human habitation within the boundaries of the Arctic National Wildlife Refuge – the Inupiat Eskimo village of Kaktovik.

KAKTOVIK

At first sight, it was clear that Max had told the truth. The oil industry *had* done well for the people in the area. Many of the houses had pickup trucks as well as snowmachines, and smooth gravelled roads led between the buildings. There was proper water and sanitation too – something which Shishmaref had notably lacked, where 'honeybuckets' (a bucket with a bag which is carried out and dumped centrally when full) had been the only toilets. Oil money had also brought the village – which, like Shishmaref, had once been just a few earth houses stuck out on the barrier island – a high school, a fire station, a police department, a community centre, a water plant, a power plant and a municipal services building.

The industry has also brought jobs to Kaktovik. Many of the young men and women work in Native-owned oilfield contracting companies, which is helping to improve the standard of living and keeping unemployment – the scourge of Native communities – down to tolerable levels. As I talked to people around the place, I rapidly got the impression that not even the elders felt nostalgic for the days when the Eskimos had lived entirely off the land. It had been a difficult existence: life expectancy had been much lower, and during the worst winters whole families had starved to death.

That's not to say that the subsistence aspect of daily life has been completely ditched: Kaktovik's annual whale hunt, carried out by the men in a flotilla of small boats, is the year's social high-point, and caribou, seals and fish are still vital parts of people's diet and culture. In fact, this conscious dependence on a clean sea leads to the one area the Eskimos *do* stand up and oppose the oil industry – in its moves towards offshore drilling. A spill under the ice would be nearly impossible to clean up, and would spell disaster for fish, whales and seals alike.

I was invited to a family house that afternoon. Jack Kayotuk was slicing up squares of beluga whale blubber in a bucket, a delicacy known as *muktuk*, when I arrived. 'Yep, it's mighty fine tasting stuff,' he said approvingly, as I chewed some of it. It tasted like fishy rubber, fatty and impossibly rich. Jack carefully peeled the grey skin off the fat and pale-pink meat (it reminded me of pulling sticky tape off a roll). There was caribou and rehydrated mashed potato to go with it. 'I've never been south of the Arctic Circle,' Jack told me with a grin. 'It gets too damn hot down there.'

I asked if he supported the oil industry.

'Yeah, and I'd like to see oil drilling in the Arctic

National Wildlife Refuge too. I think it would be all right for Alaska and for this town also. It would give us all the jobs that we need.' He mentioned how high the cost of living was in remote communities where everything had to be flown in.

Later that evening there was a knock on the door of the Waldo Arms, the homely log cabin-cum-hostel where I was staying, and Ida Angasan came in, stamping the snow from her boots. Fifty-five-year-old Ida was Administrative School Secretary, and fond of talking to visitors. I fetched her a coke, and she plonked down on the sofa in front of the television set. The local channel was broadcasting rolling text messages to all the villages, about the weather, upcoming social events and so on.

'I'm for drilling,' she declared enthusiastically. 'If they do it with safety and caution. After they drill I've seen how they put everything back together.'

I asked her why.

'The main reason is my own students – they are our future. We need a new gym, we need a new school. It's not big enough to have state championships for basketball and volleyball. I want a full-sized swimming pool too.' She laughed. 'I'm not asking for much, am I?'

'What about the wilderness?'

'I don't live in the wilderness. I'm a hundred per cent Inupiat Eskimo. This is our land. We live off the land, we subsistence hunt, we do our three whales every fall . . .'

And had she noticed any changes in the local environment?

'Oh, yeah,' Ida began. 'There's no icebergs any more. When we used to go whaling there were icebergs – we used to get fresh water from them. Then in the past few years, it's like all of a sudden . . . there's no ice. It all melts away.' She paused. 'I think it's endangering our polar bears, our seals, our ducks. I was in front of my house tonight and I saw this strange little bird – those birds come down from the mountains, so maybe it's getting warmer out there.'

She was now in full flow. 'You go to Barrow and there's open water in January. That's very unusual. This is the second year in a row that there is open water. And when it goes out, it doesn't leave chunks like it used to – it just disappears.'

I asked her how she felt about it.

'It matters to me. I don't understand it. Is it because we're not putting enough oxygen or too much pavement down, or not planting enough trees? I've seen how it floods now and gets hot in all of the US.'

Could it be global warming?

'What else could it be? I don't know.' She asked me to explain more about global warming. I told her about greenhouse gases, about the rapidly-rising temperatures, about the disproportionate effect on the Arctic north and how much worse it was likely to get. I told her what I'd heard in Fairbanks from Professor Gunter Weller, and what I'd heard from the Native residents of Shishmaref and Huslia. Her shoulders drooped as she listened.

'Well, all I can say is God bless us all,' she said quietly. 'All I know is we're in the billions now, and we all try to survive.' She sighed. 'I agree with you. If I knew more about it, I would do something. I really would.'

My time in Alaska was drawing to a close. I had expected Kaktovik to be hostile and bleak because of the oil connection, but it was just as warm and welcoming as the other Native villages. People stopped to give me a lift if I was spotted walking the few hundred yards from the hostel towards the centre of the village, and I was constantly invited into their houses as if I were an old friend.

Another thing that impressed me was the concern the Inupiat Eskimos had for their local wildlife. For a start, everyone in Kaktovik is obsessed with polar bears. People don't seem to shoot them like in Shishmaref – instead they drive to the end of the airstrip, where the whalebone dump is, and sit in their pickup trucks to watch entranced whilst the huge bears lumber around sniffing for any bits of remaining meat.

There are only 20,000 polar bears left in the world, and on my last evening in Kaktovik I was keen to see one in the wild. Travelling up to the end of the airstrip with a local hunting guide called Robert Thompson, we circled the whalebone dump, but it was empty.

Robert turned round on the snowmachine. I was standing on the sled at the back.

'Let's go a bit further afield,' he called out.

We travelled east, to a spit of land where the sea ice had piled up high against the beach, making a ridge about twenty feet high – the only vantage point for miles around.

Robert stopped the snowmachine and dismounted, moving gingerly forward, revolver cocked.

'Sometimes they can come straight at you from behind these ice mounds,' he told me. 'I don't want to take any chances.'

It was well after midnight, and the sky was cloudy, with a strange reddish light making distances difficult to judge. As I peered over the Arctic Ocean, each successive snowdrift seemed to metamorphose into a polar bear and then back again. The wind was bitter, blowing spindrift between the mounds of ice.

'There's no one between here and the North Pole,' muttered Robert, as he scanned the horizon with binoculars.

Then I saw it – a distinct yellow dot moving in the distance. 'There!'

Robert whipped round. 'Oh yeah, I got him. Quick – let's go closer.'

We leapt back onto the snowmachine and headed north. Suddenly the bear popped up right in front of us, and then – startled by the noise of the engine – quickly loped off. It stopped again two hundred yards away, the black dots of its eyes and nose amongst the yellow fur clearly visible to the naked eye. It yawned and lay down for a while, before lumbering off again at a surprisingly fast rate towards Kaktovik.

We followed it, and as we neared the village, I could see that several cars were already moving down the airstrip to marvel at the scene.

*   *   *

I felt immensely privileged to have seen a polar bear – the more so because of how threatened these beautiful animals are going to become as climate change destroys their habitat over the next few decades. Already there is evidence from Canada's Hudson Bay that polar bears are less well nourished and bring up fewer cubs in the years when sea ice breaks up earlier.[32] And this is, unfortunately, only the beginning.

With temperatures rising ever faster and sea ice coverage shrinking fast, polar bears – together with other ice-dependent animals like seals, walruses and belugas – are going to be squeezed onto a smaller and smaller remnant of floating polar ice during twenty-first-century summers.

Once that perennial ice disappears for ever – as it is likely to do within the next hundred years, according to the latest predictions[33] – the entire Arctic marine ecosystem, as we currently know it, will be destroyed. The frozen North Pole will cease to exist, leaving open water at the top of the Earth. The polar bears will have nowhere left to go, and their extinction is near certain.

This spells disaster, of course, not just for the animals but for human populations too – not just the residents of Shishmaref and Kaktovik, but all the Native people living in Canada, Siberia and Greenland – who currently depend on them. As a US government study drily points out, 'few adaptation options are likely to be available' once the animals begin to disappear for ever.[34]

Time is running out too for the land areas of the Arctic. With twenty-first-century warming predicted as high as a staggering 10°C,[35] much of the remaining permafrost is likely to thaw – further damaging forests, houses, roads and other infrastructure, and raising the spectre of massive releases of the greenhouse gases carbon dioxide and methane from bogland hitherto inert and frozen.[36]

In addition, the area of tundra is likely to decrease by two-thirds, a looming catastrophe for all the animals and plants which are adapted to this fragile Arctic ecosystem. This is ironic too, considering the current debate over oil drilling in the Arctic National Wildlife Refuge and its potential effect on the caribou. Because tundra mosses, forbs and lichens are a vital winter food source for the caribou, climate change could decimate the herds whatever the outcome of the oil drilling debate.[37]

This is just one of several bitter ironies facing Alaskans as their entire environment morphs with the rapidly-rising temperatures. The biggest and bitterest of all, of course, is that an overwhelming majority of state residents still seem deadset on pumping out their fossil fuel reserves for as long as the oil keeps flowing – whatever the eventual cost to their climate.

This dissonance sheds some light, perhaps, on the complexity of human psychology, and how difficult it is to tackle societal denial based on wilful ignorance and self-interest. But it also illustrates the wider struggle that modern civilisation in general is going to face if it is to change its ways in time to head off the worst of the looming catastrophe that lies ahead.

In this sense, the dilemma facing the residents of

America's largest and most northerly state is one which faces all of us, each time we boil a kettle, switch on a light, drive a car, or vote. It's not unique to Kaktovik, Fairbanks or Anchorage. In this modern, interconnected, energy-hungry world, we are all Alaskans.

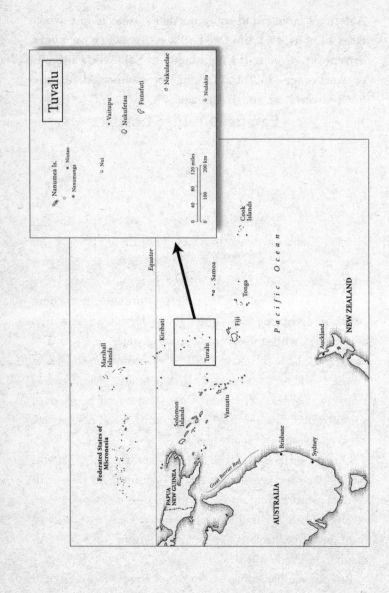

3

Pacific Paradise Lost

Nothing much happens in Tuvalu. For a while I found this charming, then it drove me crazy, and then, just as I was about to leave, I began to find it charming again. Some afternoons a tropical shower will break the heat with five minutes of torrential rain, thundering through the coconut palms and turning the dappled-blue lagoon surface into a grey mist with millions of exploding water droplets. But most of the time the temperature is so oppressive that after-noons are better spent lounging in the shade of a pandanus tree on the breezy ocean side of the island, where groups of sun-browned children play chicken with the Pacific rollers surging up onto the rocky reef.

On the surface it seems like life has tripped by at this gentle pace for centuries, and will continue to do so for centuries more. Surrounded by thousands of miles of open ocean, Tuvalu's Funafuti atoll feels the centre of its own little universe, isolated from the clamour of a rapidly-changing outside world.

But change has come to Tuvalu, change of an uninvited

and menacing nature. Bit by bit, as glaciers melt and the oceans warm, global sea levels are creeping up. Over the last half-century the rate of rise has averaged just a couple of millimetres a year, but already it's beginning to accelerate,[1] in tandem with rapidly-rising world temperatures. The minuscule increments of the past have stacked up, leading to a steady cumulative effect which is already taking its toll on island life.

For years Tuvaluan political leaders have toured the big UN conferences, pleading and cajoling industrialised country governments to reduce their greenhouse gas emissions. The Tuvaluans – together with their colleagues from Kiribati, the Maldives, Samoa and other low-lying island nations – became the symbolic first casualties of global warming, fêted on the media circuit.

But nothing much was done, and the Tuvaluan politicians, betrayed by false promises, eventually returned home empty-handed to feed their pigs and sit watching with impassive faces each year's high tide rise a little higher than the last.

And no amount of impassioned talk can change the laws of physics. Thoughout all the meetings, the press conferences and the speeches, the glaciers and ice caps kept melting and the seas – filled with this new water and the 'thermal expansion' caused by the ever-increasing warmth – kept rising.

Tuvalu's ocean clock is still ticking today, but it's nearly out of time. The people of the islands are now faced with the choice they've always dreaded – to move, and live cultureless and uprooted in a foreign country, or stay on the land of their forefathers and die. From the distant vantage point of my home in Oxford, I heard that the choice had finally been made.

FUNAFUTI ATOLL

The first person I met in Tuvalu was Paani Laupepa, the tall, solidly-built Environment Ministry official long one of the most articulate voices of his country's plight. I'd seen him quoted in countless media reports, and was looking forward to questioning him more.

'No, no, no,' he insisted. 'You must go and rest. We can talk any time – you go and lie down.' Then he beetled off on his motorbike – everyone on Funafuti has motorbikes – and left me little option but to obey.

It was ridiculously hot, and I lay under a mosquito net whilst a small fan whirred impotently a few feet away. Already Suva, the Fijian capital where I'd boarded my flight a thousand kilometres away over the cloud-flecked open Pacific, seemed like a different world. As I lay on the bed, sweat quietly dripping, I could hear nothing, just the occasional buzz of a passing motorbike, the swish of wind in the palms above, and the far-off rumble of the ocean.

Once the harsh sunlight began to soften a little, I wandered outside to explore. A hundred metres on my left was the lagoon, fringed by a narrow beach, the water mottled with purples and light blues where the sea floor alternated between sand and rock. A few women stood chatting in the water, only their heads showing above the rippled surface –

looking as natural as old ladies passing the time of day at a London bus stop. Every now and then someone would heave themselves out of the sea fully-clothed, and set off, dripping, back to their house. I marvelled at their almost amphibious lifestyle – being wet or dry made little difference in this equatorial heat.

The ocean was little more than five minutes away on the outer edge of the island. From the air, Funafuti looks like a teardrop, with a scattering of small islands strung out along the narrow atoll rim, itself broken by long sections of submerged reef lying between the calm central lagoon and the open ocean. Looking out over the lagoon from the village, you can see the curved line of neighbouring islands stretching in both directions, their coconut palms standing like tiny upturned pins where the two halves of the teardrop meet on the horizon.

In places the strips of island are only a few metres from one side to the other, and at its widest – where the town and airstrip are built – Funafuti atoll is little more than five hundred metres across. In fact the whole country of nine tiny islands has a land area of only twenty-six square kilometres – a tenth the size of Washington DC. I wondered how it must feel to be stuck there with a hurricane bearing down on you, and shuddered.

I still didn't know whether the rumours of imminent evacuation were true. For this reason, I went to visit Tuvalu's top civil servant Panapasi Nelesone who, as Secretary to Government, was the man most likely to have an answer.

Thick-set and bespectacled, Panapasi showed me into the Prime Minister's office (the PM himself was at a Commonwealth meeting in Australia), so that we could talk in private. Then he sat down in the Prime Minister's chair, leaned back, and emitted a long, laboured sigh.

It was true, he said. Tuvalu was indeed preparing for the end.

I looked around the office, thinking of what to say next. Some optimistic building plans, together with an intricate model ship, lay on the Prime Minister's desk. An air-conditioning machine hummed outside the yellow one-storey building that housed most of Tuvalu's government.

'We couldn't just sit back and do nothing,' he explained, in a soft, halting tone that belied his official status. 'And so far we have received approval from New Zealand to allow seventy-five people a year to go there.'

'When does the agreement start?'

'We don't know, but it will be this year. We will try to conclude everything this year and allow people to migrate.' He paused, before revealing that initially the quota had been set at three hundred people a year – but with a population of only ten thousand, losing so many would have left the islands depleted too soon, their society fragmented and government services run down. So they'd settled on a more reasonable figure, and would be urging people like doctors, teachers and civil servants to stay longer.

'We cannot just leave these people without health services, education services, or communication among themselves. Our islands are scattered: we have to provide services for our people.'

I still had trouble believing what I was hearing. 'But surely it must be a terrible thing for people to face having to leave their home?'

Panapasi sighed again, and continued in the same soft tone. 'Exactly. We don't know what will happen in the future. We may lose our culture. We may lose our identity as Tuvaluans. It will take time for our people to accept that, once we're in another country. For a population like Tuvalu to move and leave this place, that would be something hard for our people to finally accept.'

'What about you and your family?'

He waved my question away. 'No, no, I have no plans at this stage to move. I was given overseas training by my government, and I think I have a duty to help my country now.'

As we finished the conversation, I made the mistake of using the word 'evacuation'. He broke in sharply: 'It's not an evacuation. We have not yet reached the stage where we must evacuate people. We know there is the threat of global warming, and the government doesn't want to sit back and do nothing. So this is a migration programme, a gradual kind of thing over time, not an evacuation as such, where we have to immediately move people.'

Back outside, a strong breeze had risen whilst I was talking to Panapasi. The lagoon surface had taken on a ruffled, dark appearance, and ominous clouds were building up.

The year's highest tides were forecast for the next couple of days, and already the water level was only a foot or so below the edge of the dirt track behind the government offices. Choppy waves occasionally broke over the edge, washing fallen palm leaves up onto the dry land.

As I walked past the lagoon I reflected on whether Panapasi's refusal to use the word 'evacuation' was wise. On the one hand, it would calm people, prevent panic, and it was important not to undermine a functioning society amongst those people who chose to stay in Tuvalu for the foreseeable future.

But it also seemed to understate the urgency of Tuvalu's crisis, potentially making it easier for the countries responsible for causing global warming to continue evading their responsibilities. Australia particularly, under the far-right Howard government, seemed intent on doing just that: not only had it turned down the Tuvaluan government's requests for discussions about the 'migration' issue, it was also refusing even to ratify the very limited Kyoto Protocol on climate change.

By this time I had reached a grassy area at the top of the airstrip, where a well-attended football game was underway. Upwards of fifty men and boys were leaping and chasing the ball around, all just a few yards from the ocean. Between them and the waves, forming one edge of the football pitch, lay a three-metre-high bank of rubble – the highest bit of land on the island – pushed up off the beach by Hurricane Bebe in 1972.

This storm was so devastating that 21 October is still

remembered every year as 'Hurricane Day'. I later heard the bank of rubble referred to by Paani as 'Mount Howard', after the Australian Prime Minister, and someone in the Environment Ministry was apparently even having a T-shirt printed with the slogan 'Tuvalu Mountain Rescue Team'. Humour often comes with adversity, and Tuvalu has plenty of both.

I climbed Mount Howard and sat watching the ocean breakers roll and crash on the reef. A luxurious glow began to light up the whole scene as the sinking sun painted distant storm clouds orange, red and then crimson, the sky reflecting in a shimmering light off the breaking Pacific swells. The air was so clear that the horizon stood out as a stark blue line, with some of the clouds so far away that only their tops were still visible over the curvature of the Earth itself.

There was a party that night, as there is every night with a suitable excuse on Tuvalu, to which Paani had invited me. All I had to do was follow the echo of the Tuvaluan drums towards the large *falekaupule*, an open-walled thatch building that doubled as the country's parliament in daytime and main party venue at night. They weren't drums exactly; instead it was more of a low table, which the biggest men – on the ground around it – beat in a heavy rhythm with their strong hands. Around them the dancers moved in perfect time, everyone wearing pandanus-leaf skirts, the women in delicate white-and-red blouses, and the men bare-chested except for a few casually-draped banana leaves.

It would have been nice, I thought absent-mindedly, to have visited Tuvalu before the missionaries first arrived, when everyone danced completely bare-chested without a second thought. Some missionaries had even tried to ban traditional dancing altogether, hoping presumably to replace it with self-flagellation and sexual guilt.

I was quickly spotted and ushered forward to a row of VIP chairs at the front. Almost everyone else was sitting cross-legged on mats on the floor but, thankfully, the hospitable Tuvaluans have realised that all *palagi* (white men) are so chronically stiff-limbed and unfit that we are unable to sit cross-legged for more than twenty minutes. I noted with interest that the visiting Japanese dignitaries – in whose honour the party was being held, thanks to their government's donation of a new inter-island ship for Tuvalu – suffered no such problems, and sat contentedly cross-legged on the floor as the chorus of singing rose and ebbed around them.

A distinguished-looking old man, his thick shock of black hair with just a trace of white around the temples, sat down next to me and began, in a low, growling whisper, to translate the songs. I could see that people respectfully moved aside for him, though at that point I had no idea who he was.

'This is about *alofa*, meaning love or presence in our Tuvaluan language,' he spoke in perfect, well-educated English. 'What they are dancing for is the beauty of the islands. We all live under one hut and come together.' He paused, as the dancers swirled and shook in front of us.

'This is the *fatele*,' he went on. 'It means "make plenty": plenty of movement, plenty of happiness. It is a peaceful dance. We don't dance with weapons – we dance with our hands and we dance with our hearts.'

I bumped into Paani (who had been master of ceremonies at the celebration) afterwards, and asked him who my neighbour had been. 'The gentleman you were sitting next to? That was Toaripi Lauti, first Prime Minister of Tuvalu after we became independent in 1978. You should talk to him in the morning – he has a lot of knowledge and a lot of experience too.'

The appearance of normal life on Funafuti confused me. I knew that sea level rise was a decades-long process, and that Tuvalu's evacuation would likely never mean a queue of people standing desperately next to the airstrip waiting for the next plane out – unless, perhaps, another cyclone was on the way.

But why was the road crew still driving around in their bulldozer the following morning, laying tarmac on Funafuti's dirt tracks? And why the plans to build a new government office on land which would soon be underwater? I wondered how people could continue to believe in their future as Tuvaluans, when they must know deep down they have none. I went to find Toaripi Lauti.

I discovered very quickly, however, that Toaripi wasn't the sort to mope. He was upstairs in his Town Council office, where since being Prime Minister, and subsequently Governor-General, he had been trying to live out his retirement usefully.

'I always have bare feet,' he said as we shook hands, noticing me looking at the floor. 'Haven't worn shoes in years.' Reaching behind his desk, he brandished a framed photograph at me, showing him with the then British Prime Minister, Margaret Thatcher.

I couldn't hide a grimace. 'You met Thatcher?'

'She was very nice,' he rebuked, grinning. There followed several long stories about his attendance at Charles and Diana's wedding ('It was wonderful'), his own private conversation with Diana at the time ('Very nice lady, beautiful') and Charles's visit to Tuvalu not long after independence ('He was very good, a young fellow'), as well as the time when some unsuspecting American yachties met him on the beach and asked him to do their laundry, without realising that they were speaking to the country's Governor-General (he happily obliged). I eventually steered the conversation round to Tuvalu's current problems, and asked whether he too believed that the sea level was rising.

'I am one of those people who work in our traditional gardens,' he began. 'These are in sunken pits, where we dig down to plant *pulaka*, that's something like taro, about a foot below ground level. That was fine, I used to go and harvest them after a few months without any problems. But then about ten years ago I found that the bottom part of these plants was rotting.'

He looked out of the window, where a shower of rain had

just fallen, dampening the midday heat. 'And not only that – I tested all the other ones, and they were the same, all the same. I found that sea water had been coming in with the high tides, making it too salty for the plants to grow. And that was the first time I knew it was happening.'

Armed with his own experience of global warming's first effects, Toaripi had led Tuvalu's government delegation to the UN climate conference in Kyoto (which resulted in the agreement for a Kyoto Protocol) back in 1997. There he gave a speech, not just about the salt water intruding into his *pulaka* crop, but about another piece of evidence fitting the whole puzzle together.

'I told them that another of the things proving that our water has risen was that one of our islands was completely washed out. No coconut trees, no living trees, no plants were left – only the base of it, just rocks and sand.'

'What was the island called?'

'Tepuka Savilivili.' I wrote the name down – it sounded well worth a visit.

'And in Kyoto,' Toaripi went on, 'the scientists were all there explaining about these gases, these emissions and all that. So I said, well, if that is what is causing this, they should do something to stop it.'

But no one had. And so it had come to this – to 'migration', to the extinction of an entire island country and its way of life.

Toaripi's mood was sombre, but there was a determined look in his eyes.

'My thinking is that now is the time for preparing a place

so that when people move they can move with their traditions, their customs and their culture. Some people say no, no it won't happen – they don't believe in it. So I say, well, which one would you like – would you like to stay here and then every one of us will die and there will be no more Tuvaluans? Or that we prepare and move to another place where we can survive?'

The determined look was still in his eyes, and I could see that it was born from a lifetime of difficult decisions.

'But what about you personally, you and your family?' I asked. 'What are you planning for the future?'

'I want my children to be safe,' he answered evenly. 'I tell them: you leave so that Tuvaluans will still be living, with their younger ones growing up.' He smiled sadly. 'But I want to stay on this island, you know. I will go down with Tuvalu. This is my thinking.'

'So younger Tuvaluans should move elsewhere, but you'll go down with the ship?'

'That is what I am thinking.'

HIGH TIDE

Early that evening, next to the airstrip, a puddle appeared. It wasn't raining; in fact there was hardly a cloud in the sky. But the puddle kept growing, and as I watched, it was gradually joined by other smaller puddles, which appeared, as if by magic, at the side of the tarmac. I splashed to the centre of one. Clear water was welling up through a small hole in the ground, just like a spring. Trails of bubbles were surfacing from other small gaps, as water replaced air somewhere deep underground. I stuck my finger in one of the holes, and cool water welled up around it.

'It's the high tide,' announced Paani, who had suddenly arrived on his motorbike. Tuvalu's coral rock was hollow, he explained. So as the sea level rose around it, underground water was forced up onto the surface. That was the reason, he continued sadly, that no sea wall could protect Funafuti atoll. The islands would flood from the inside out.

'What's it going to be like in ten years' time?' he asked. 'It's going to be right up on the airstrip. We're going to be left stranded.' He clicked his tongue. 'It's no good.' Then he kicked the motorbike back into life, and rode off to inspect the rest of the island.

On the near side of the airstrip, the back gardens of several houses were beginning to submerge. One of them (ironically, the office of Tuvalu's Immigration Department) had water welling up through holes right around it, which gathered in a large pool under a coconut tree. The pool began to lap up against the Immigration Department's front step. Two girls came out of the building and edged around the flood, trying to keep their smart shoes dry.

'Do you believe the sea level is rising?' I asked them.

'No,' they giggled. 'We're Christians. God will protect the island.'

Two older women were standing nearby, one with a small child in her arms. They both looked worried.

'Since the 1990s there has been a big change and you can see that the water is really rising,' the one with the child told me. 'When I was a kid, there wasn't any water around the airstrip – now look at it.'

Her friend nodded. 'I've noticed that the sea was becoming higher than it was in the past. We never saw that before.' She looked down at the ground. 'It's becoming scary – we don't know what will happen in the next few years.'

At the top end of the airstrip the water had reached nearly a foot deep. Whilst I waded around, a kid swam past, using an old piece of polystyrene as a float. Fish darted around my ankles in the floodwaters, dark shapes investigating their new terrain. The flooded area was now over a hundred metres across, and encompassed much of the central area of Funafuti atoll.

Outside a large communal building three young men were standing around a barbecue – the water now nearly up to their knees – frying chicken and fish for a funeral reception. A procession of women joined them, balancing big bowls of food on their shoulders with one hand whilst hitching up their skirts with the other to stay dry as they waded through.

Panapasi was on the porch of one of the nearby houses surveying the scene. 'Maybe we should try growing our gardens hydroponically,' he joked bitterly.

I drove back down the airstrip to the southern end of the island, where a row of houses which backed straight onto the lagoon seemed to be bearing the brunt of the high tide. Waves were breaking over a makeshift sea wall, soaking the squawking chickens in a nearby coop. A plump mother sat next to it, washing clothes and keeping an eye on a group of kids as they played in the water. All of them – mother and children together – would periodically disappear in a mass of flying foam and spray each time one of the bigger waves hit the wall. I beat a hasty retreat back to the bike with my camera and notebook dripping salty water.

Back at the government offices, the waves were breaking right onto the track, surging between the coconut trees that usually marked the top of the beach. But I could tell, from the scattered debris up in front of the Environment Ministry, that the high tide was already beginning to subside. Tuvalu had got a reprieve – this time at least.

The following morning I finally got Paani to sit down and give me a proper interview. Dressed in a smart grey *sulu*, he radiated quiet confidence as we walked down from the government offices to the beach. I knew from talking to others that he was one of the most respected officials in the government, and that – through a succession of Prime Ministers – he had taken consistent charge of bringing Tuvalu's concerns to the attention of the outside world.

He teased me affectionately about my sensitive white skin when I insisted on wearing a floppy hat to keep the sun off my face. The sea had come up again in the early hours, but now everything seemed to have returned to normal. Only small waves were coming in off the lagoon, curling harmlessly against the pebbles.

We began by discussing the floods. 'It's becoming more extensive,' he told me bluntly. 'Twenty years ago it also happened, but on a very small scale. Now we see a much higher percentage of low-lying areas being inundated.' He also told me about how saltwater intrusion was damaging the *pulaka* pits, and promised to show me an example that afternoon.

But, unlike most other Tuvaluans I'd met, Paani also wanted to talk politics. Specifically, he wanted to talk about how his country – with its minuscule contribution to the global warming problem – was about to be its first casualty.

'We are being made the victims of something that has nothing to do with us at all,' he complained. 'The industrialised countries caused the problem, but we are suffering the consequences. We are on the front line of climate change through no fault of our own, and it is only fair that people in industrialised nations and industries take responsibility for the actions they are causing. It's the polluter pays principle – you pollute, you pay.'

The figures support Paani's argument. Tuvaluans do produce greenhouse gas emissions (mainly from motorbikes, boats and diesel-generated electricity), but in tiny quantities compared to Western countries. The average Briton produces twenty times more than an average Tuvaluan, for example, and the average Australian thirty times more.[2] And in any case, Paani went on, Tuvalu was now planning to move to a carbon-neutral economy, phasing out the use of fossil fuels in power generation altogether.

Because of this monumental unfairness, Tuvalu was now planning legal action against countries and industries that refused to take action to reduce their greenhouse gas emissions. Two countries in particular stood out: the United States and Australia.

Paani Laupepa wasn't the sort of man to mince his words. 'To us, as Tuvaluans and people living on low-lying atolls, America's refusal to sign the Kyoto Protocol is like an attack on our freedom and democratic values, because it is affecting the entire security and freedom of future generations of Tuvaluans.'

And Australia?

'They slammed the door in our face. We believe they should be helping us here, as a fellow Pacific country, not following the United States.'

In fact, not only had Australia spurned Tuvalu's request for help in resettling some of the people made homeless by global warming, but it then went on to ask Tuvalu to host a prison camp for its own unwanted asylum-seekers. Hugely offended, the Tuvaluans turned the request down, and the Howard government eventually set up its camp on remote islands in cash-strapped Nauru and Papua New Guinea. (Many Tuvaluans detected a hint of racism in Australia's 'Pacific Solution' to asylum-seekers, and – perhaps antici-pating their own looming refugee status – also felt a degree of sympathy with the dispossessed boat people.)

So what was Tuvalu hoping to gain through the courts?

'We want to see governments take serious domestic action to reduce emissions,' Paani told me. 'Certainly we will be seeking compensation for the problems we are facing. This is one of the biggest threats that has ever faced our nation, and I think the entire world.'

But the compensation issue also raised problems.

'How do you put a price on a whole nation being relocated? How do you value a culture that is being wiped out? How do you put a price on ancestral homes which are being destroyed?'

I had no answer, and I suspected that any international court would have a tough time answering it too. Indeed, it seemed almost inconceivable that any monetary compensation could make up for the losses that he was describing.

And in the meantime, well before any long-term evacuation was complete, the impacts would keep on mounting. Paani showed me one of the freshwater *pulaka* pits which had been affected by rising sea levels. The huge waxy leaves of the plants had yellowed, and many of the younger plants were spindly and dying. It was difficult to imagine how the area had looked when the *pulaka* was thriving, its valuable starchy tubers expanding underground in readiness for an approaching feast.

A gardener was squelching through the mud, tending to the few plants which remained healthy. He didn't hold out much hope. 'It was a very high tide this month, so in the next two weeks these plants will change to yellow too. Three-quarters of the plants in this area have died,' he said forlornly.

So what else did people eat, if they couldn't grow their own food?

'Now people have to eat imported food, like rice,' he replied. 'So if we don't get enough money we can't eat very well.'

Saltwater intrusion has destroyed so many *pulaka* pits that this once staple food has become a luxury – served mainly at ceremonial occasions rather than eaten in the home. In Tuvalu's outer islands, which still maintain a subsistence-based traditional diet, the problem is even worse, especially as imported food costs money that many don't have, and the supply ships can be held up for weeks at a time because of bad weather. On Vaitupu, north of Funafuti and Paani's home island, several *pulaka* pits have been abandoned – to be reclaimed by the coconut crabs and the mosquitoes.

In the Tuvaluan language, a person without land is known as *fakaalofa* – literally 'deserving of pity' – a status to which all islanders may eventually find themselves reduced in decades to come.

There was no party that evening, so I went to visit Mataio Tekinene, an official in Paani's Environment Ministry, at his ocean-front house. Mataio, who looked half-Indian with his smooth, round face, was in his element, wearing only a *sulu* wrapped around his waist. The sun had just set, and a full moon was lifting itself above the clear horizon, hovering above a layer of silvery blue clouds.

Mataio was watching a group of children playing in the surf. Some of them were his, some of them weren't. I never found out which were which, and it didn't seem to matter. Children in Funafuti are everyone's responsibility, and roam around with a careless freedom that most British kids could only dream about.

Mataio's small brick house was set back a short distance from the ocean, but most of the family life seemed to be taking place in the kitchen-cum-sleeping area – a metre-high wooden platform with just a tin roof, covered with sleeping mats, pillows, pots and pans, mosquito nets and surrounded by lines of washing. The platform was slung up through the pandanus trees that lined the top of the beach, and Mataio's wife sat cross-legged inside, preparing food. He and I perched on a dugout canoe, and looked out to sea.

The normality of life on the island was still confusing me. 'I don't understand it,' I began. 'Why build roads and put up streetlights at the same time as planning to move people off the island?'

He chuckled. 'Seems like a waste of money, eh?'

'Well . . .'

Mataio shrugged. 'Development cannot stop. People will be here for many years yet, and we need these services.' He said nothing for a while. 'But people are already leaving, it's true. Lots of Tuvaluans now have relatives living in Auckland. Some go for employment, or maybe education. People think there's no future here, so they leave.'

The children's game showed no sign of ending, despite the fading light. Between swells they all ran together towards the sea, scampering over the rocky reef. Then a huge wave broke over them and they emerged from the foam howling with glee and rubbing the warm salt water from their eyes.

Mataio explained how he supplemented the family

income by fishing, taking the dugout canoe out onto the lagoon to look for rock cod, red snapper, and other shallow-water fish – in fact he'd been out between 4 and 10 a.m. that morning, before going in to work at the government office.

The tide was coming in again, and the largest waves came right up the beach almost as far as the edge of the trees. The roar of the surf, and the clatter of thousands of rolling pebbles as each wave retreated, almost drowned out our conversation.

I raised my voice. 'Have you noticed the water getting higher?'

He nodded, and picked up a stone, pitching it twenty feet away down the beach, where it was instantly consumed by a breaking wave. 'That was the top of the beach when we first moved to this spot twenty years ago,' he shouted back. 'Two years ago during a high wind the waves came right up to the house. Maybe we'll have to move again soon.'

To New Zealand?

'I think I prefer it here. The way we live, it's an easy life. In New Zealand you have to work all hours, pay the rent, and all that.' We walked back together to the outdoor platform, where his wife was serving up a meal of fish and rice. A few people were stretched out sleeping. Mataio grinned and indicated in their direction: 'It's the Tuvaluan way. Wherever you lay down, that's where you feel comfortable.'

We ate the fish raw. It was as exquisite as Japanese sushi, and available in much larger helpings, thanks to Mataio's nocturnal excursion. We also shared a couple of beers, and as the evening drew on I decided that I had become very fond of Tuvalu.

The following morning I chugged down to the Environment Ministry on a hired motorbike to see if anyone was around. They weren't. I checked my email on Mataio's laptop, vaguely irritated at the messages from abroad, as if they were unwelcome intrusions into my new carefree island existence. I noticed on the way out that someone had hung a freshly-caught tuna from the wooden frame of the porch, where it swung in the breeze next to the dripping air-conditioning outlet.

Then, increasingly bored, I went off to snorkel, paddling out over the near sandbanks in the lagoon to some rather forlorn bits of coral reef. Various people had warned me about sharks, so I tried to keep an eye out in all directions through the cloudy water. I was surprised by a whole school of eagle rays, which swooped around me like enormous flapping birds, before turning and wheeling off into the murk with frightening speed. An old man paddled past in a canoe.

'*Talofa*,' I said, proud to use my one word of Tuvaluan.

'*Talofa*,' he replied, with a smile.

'Fishing?' I was idly treading water.

'No – going to feed the pigs.' I noticed the coconuts in his canoe and nodded.

He paddled slowly off, and I went back to the shore, lay down on the beach – getting into the Tuvaluan way – and closed my eyes. I was woken by a rainstorm somewhat later on, and went back to lie in the lagoon, which was as warm as bathwater against the cold drops of rain.

Tuvalu may turn out to be one of the shortest-lived countries in history. Although thought to have been settled for at least two thousand years, it has only been a nation in the modern sense since 1978, when the Union Jack was lowered over Funafuti atoll by the colonial commissioner (Princess Margaret was indisposed). Before that, it was – in union with the modern-day Kiribati – the colony of the Gilbert and Ellice Islands, and had been since 1892 when the British first arrived (much to the jubilation, or so they claimed, of the islanders), in order to prevent the Germans getting there first.

One of the new colony's Resident Commissioners was Arthur Grimble, later famous as the author of *A Pattern of Islands*, which must qualify as one of the earliest and best expositions of the 'bumbling white man makes a fool of himself in front of the natives' school of travel-writing.[3]

The following passage, in which a pasty-faced twenty-five-year-old Grimble is introduced to the correct way to drink from a coconut, is one of my favourites. Supervised by a seven-year-old girl, he has just swigged the whole thing, and handed the empty shell back with a curt 'thankyou', leaving the little girl shocked at his bad manners. 'See, this is how you should have done it,' she lectures him sternly.

She held the nut towards me with both hands, her earnest eyes fixed on mine, and gave vent to a belch so resonant that it seemed to shake her elfin form from stem to stern.

'That,' she finished, 'is *our* idea of good manners.'

Curiously, and perhaps a little dispiritingly, Grimble is not remembered as fondly in Tuvalu as a reader of *A Pattern of Islands* might expect. In an official history, Grimble sounds something of a sourpuss: '[He was] a conceited and ambitious man, who believed he knew better than anyone else what was best for his subjects. Like many colonial officials, he regarded them as children.'[4]

Historical evidence seems to be on the side of the Tuvaluans, given that Grimble is on record as having introduced a host of patronising, often pointless rules and regulations which were only rescinded after many years of complaint and protest. These included the requirements that all persons spend Fridays cultivating their trees and gardens, that islanders refrain altogether from inter-island sailing, and silliest of all – that a bell should be rung every evening at 9 p.m., after which anyone visiting the latrines was required by law to carry a lamp.

Perhaps Grimble was only following the example set by the first foreign missionaries, who instituted fines for the offences of fighting, dancing, tatooing, telling lies, and failure to observe the sabbath. Offences committed on the sabbath always carried a much higher penalty (perhaps on the grounds that God was more likely to be watching). On the Tuvaluan island of Nukufetau, for example, the crime of fornication carried a fine of 300 coconuts, rising to a rather heavy 1300 coconuts if committed on a Sunday.[5]

Before the missionaries arrived and informed everyone that the islands had in fact been created by the Christian God, Tuvaluans believed that their nine atolls had come into existence during a fight between a flounder and an eel. The two fish, once the best of friends, had decided to test their strength by carrying a huge stone. The competition had led them into a fight, during which the flounder was crushed under the stone and flattened. After suffering a heavy blow to the stomach, the eel vomited so much he became long and thin, rather like Tuvalu's coconut trees, whilst the flounder's body became the model for the islands themselves. To this day, it's still *tapu* (taboo) to eat eel.

The scientific origins of coral atolls were a topic of fierce debate during the nineteenth century, which was resolved in typically far-sighted style by Charles Darwin, who proposed that these circular islands had been built up over millennia by shallow-water coral growing on the rims of slowly-subsiding volcanoes. He was proven right fifty years later when a Royal Society of London expedition drilled rock samples to a depth of 300 metres on Funafuti itself, and found that they did indeed include the fossilised remains of shallow-water corals – which had been buried as their volcanic rock base subsided, just as Darwin suggested.

This issue is still highly relevant, because it is theoretically conceivable that the coral reefs surrounding atolls could grow upwards and keep pace with sea level rise, keeping the islands themselves above water even in a warming world. Unfortunately, it seems that sea level rise is happening too quickly, and moreover that tropical corals are anyway in no state to put on a sudden growth spurt. Not only are they increasingly polluted and degraded, but rising sea temperatures have led to a new threat: coral bleaching.

Coral reefs are the most biodiverse marine ecosystems on the planet, containing up to nine million different types of plants and animals, including a quarter of all known sea fish.[6] They are also very fragile, and tropical coral reefs are particularly vulnerable to high sea surface temperatures, which trigger episodes of 'bleaching' – an automatic response where the coral polyp (an animal) loses its symbiotic algae (a plant called zooxanthellae).

When bleaching happens, whole swathes of colourful reef turn bone-white and rapidly die. Dive on a bleached reef a few months later, and the whiteness is gone too: instead the reef has been buried under a choking shroud of grey algae. Since coral polyps and their calcium carbonate skeletons are the foundation of the entire ecosystem, fish, molluscs and countless other species, unable to survive in this colourless graveyard, rapidly disappear too. Bleached corals can sometimes recover, but their biodiversity is reduced each time they are hit, and some reefs – especially those which are already suffering direct impacts from pollution or dynamite fishing, or where high temperatures are sustained for long periods – are killed off for good.

Coral bleaching was almost unknown until the late 1970s, when global warming began to push tropical ocean temperatures above reef tolerance levels for the first time. Since then bleaching has become an ever more regular occurrence, reaching catastrophic levels in the 1998 El Niño, when a sixth of the entire tropical coral reef ecosystem on the planet was destroyed.[7] Massive mortality was recorded in reefs around Sri Lanka, the Maldives and across the Indian Ocean, with death rates as high as 90% in some areas.[8] On the Great Barrier Reef massive corals up to 700 years old died off, showing the unique nature of the event. Yet disasters on this scale are likely to become commonplace within a mere two decades, and within thirty to fifty years severe bleaching like this could be an annual occurrence.[9]

As many scientists have pointed out, there is no way that reefs can survive annual bleaching events on the scale of 1998. This leaves room for only one conclusion: that the vast majority of tropical reefs will disappear almost in their entirety from tropical seas within the next half-century – perhaps the single worst biodiversity disaster that humanity will ever have witnessed (and caused). People I met in Fiji were keenly enouraging their children to dive and snorkel on the country's beautiful reefs, knowing that they may be the last generation able to do so.

It's clear from this that Tuvalu's people can't look to their reefs for salvation from the rising sea, and in places whilst snorkelling on the ocean side of Funafuti I spotted small bleached sections amongst branching and flat corals: the first indications of their likely fate.

That evening I took a fishing boat a few miles around Funafuti atoll's rim, to a smaller island called Amatuku, where I'd heard that one of the oldest colonial buildings, dating back to 1905, was still standing but had begun to flood – a further indication that sea levels are higher than ever before.

Amatuku is almost entirely occupied by the Tuvalu Maritime Training Institute. The grounds were grassy and neat. Stone-lined paths led between a parade ground and dormitory block. Chief Engineer Delai Vakasilinki came out to meet me, impeccably dressed in a starched white uniform and peaked cap, his air of naval authority making me feel immediately rather deferential.

'Mind your head,' he advised, noticing that I was standing under a heavily-laden coconut tree. 'If one of those things hits you, you know about it.' Bottles were tied high up in the tree next to the nuts, draining off sap into coconut toddy. This sweet liquid is so nutritious that it was once used as a substitute for breast-milk, and is also the raw material for the severely alcoholic 'sour toddy'. Unfortunately everybody on the main island seemed only to drink beer, which meant I wasn't able to try out sour toddy's legendary hangover-inducing properties.

The old colonial building was a one-roomed, stone-walled hut, topped with a thatched roof. Whether Arthur Grimble ever stayed there isn't recorded – but I suspect he might have demanded something a little more upmarket.

The whole area looked perfectly normal and dry to me. We stood around and chatted for a little while; Chief Engineer Delai occasionally casting a schoolmasterly glance over to where two teams of sailor-suited trainees were playing an energetic game of football. Although it was all pleasant enough, I was beginning to suspect that I might have wasted my time.

Then the first little spring appeared. 'This is just the beginning,' said Delai, with a curt nod.

The first spring was quickly joined by another, and another, and within twenty minutes a small river was flowing down towards the old hut and collecting in a pool around it. Nor was this the only area of the island to be affected – a hundred yards away one of the coconut groves was going under. I splashed over to take a look. Millions of ants were evacuating their nests and streaming up the trees to safety, closely followed by dozens of yellow-and-black-striped lizards.

'This now happens five or six times a year during the spring tide,' Delai announced when I returned. 'It is very worrying, and long-serving staff members have said it is getting worse.' He climbed up onto a raised concrete basketball court in order to keep his feet dry. 'We can't stand it very long. If it keeps going on this way, how long can this place be safe?'

Over on the football ground, the trainees' game was becoming increasingly muddy. As one half of the pitch began to submerge, great jets of water flew up into the air each time one of the players kicked the ball.

Delai watched a tongue of water lick over onto the basketball court. His feet were going to get wet after all. I had given up keeping dry, and waded over to the old hut. The flood was knee-deep in the doorway, and a foot deep inside. An old steel bed stood out from the water, its mattress having been removed earlier to higher ground.

An hour after the first springs appeared, the football game had been reluctantly abandoned and the new lake was at least a hundred metres long. Just then there was a roar, and on the ocean side of the submerged football ground a large wave washed right up over the mound of rubble that lined the beach, throwing a torrent of foamy water down into the new lake. Amatuku was under assault from all sides.

A whistle blew, and we all stood to attention as the Tuvaluan flag was taken down over the parade ground, carefully folded by one of the trainees, and marched off for night-time safekeeping.

Although doubtless a regular daily ritual for the young sailors, it was a poignant moment – as we stood on an island lit by the setting sun, the symbolism of the lowering flag and the rising waters was striking.

Delai waved goodbye with traditional island generosity, loading the small boat with armfuls of green coconuts. The boatman opened them with a large knife once we were on our way, and we shared them together as the last rays of the sun glanced off the calm waters of the lagoon.

Of course, sea level rise hasn't just affected Tuvalu. Its effects are now identifiable in just about every coastal area of the world. Over 70% of the world's sandy shorelines are retreating, and saltwater intrusion has been documented in low-lying deltas as far afield as China's Yangtze River and Australia's Mary River.[10] On the US East Coast, where a rising sea is compounded by the gradual sinking of the land, 80–90% of beaches are eroding. In Chesapeake Bay islands mapped by settlers in the seventeenth century have split into a few tiny remnants, and the Blackwater National Wildlife Refuge, an important coastal wetland for geese and ducks, has seen over a third of its total marsh area (2000 hectares) submerged between 1938 and 1988.

Famous beaches have been affected too: at the popular resort of Ocean City, Maryland, municipal authorities are forced to truck in vast amounts of sand to counteract a beach erosion rate of five metres a decade, and the same problem is affecting South Beach in Miami. Saltwater intrusion has been documented in the low-lying Marshall Islands – another atoll nation – and in the 607-island Federated States of Micronesia, where the capital's main road is already being attacked by erosion on the beach side.[11] Britain is also feeling the effects, and official government policy has moved from building coastal dykes to a strategy of 'managed realignment' along much of the flat east coast.[12]

Before arriving in Tuvalu, I'd interviewed Professor Patrick Nunn, an expert in ocean geoscience from the University of the South Pacific, who confirmed that the impacts of sea level rise are visible right across the Pacific. On some islands hundreds of metres of coastal flats have been inundated by the gradually rising waters, leaving seaside villages cramped at the bottom of cliffs. Low-lying atolls are particularly hard-hit, but mountainous islands are suffering too: Nunn has personally identified places in Fiji, Vanuatu, the Solomon Islands, the Cook Islands, Samoa and Tonga where the rate of beach erosion is currently up to a metre per year.

I also saw some of this for myself in Vanuatu, taking a day-long tour of the main island with local meteorologist Nelson Rarua. There was evidence of severe coastal erosion almost everywhere: in many places large well-established trees – once secure on dry land – were being undercut by waves, and some had toppled over. Large sections of coconut plantation had been washed away within their owners' lifetimes. In one village several families had been forced to move their houses back from the advancing seas, and the well had been turned brackish by a saltwater flood the previous year.

A little further round the coast a headstone – 'June 1757' still visible carved into it – had been left stranded at the top of a beach, the bones and the whole grave underneath swept out to sea during a cyclone. Day by day these changes were minuscule, Nelson Rarua pointed out. But in the long term . . . he left the question hanging, pointing behind him to a poster on the wall of his office. 'Climate change: killing our islands softly,' it said.

Back in Fiji, Professor Nunn was also making some dire predictions. 'I think there are islands in Tuvalu and Kiribati and places like that where you could sustain life as it is for maybe another fifteen to thirty years,' he told me. 'I think that many low-lying coastal areas, be they in atoll countries or in high island countries like Fiji and Vanuatu, are going to disappear. In fifty years' time the geography of the Pacific region will be quite different to the way it is today.'

Again, it's not just the Pacific which is affected. Indeed, the century ahead looks grim for coastal dwellers around the world. The rate of sea level rise is likely to accelerate by around two to four times the current rate – leading to a further rise of anything up to a metre by the end of the twenty-first century.[13] If this happens, 15% of agricultural land in the Nile Delta will be inundated, displacing six million people. Thirteen million more in Bangladesh are also at risk, with nearly a fifth of that country's best rice-growing areas likely to be abandoned. In total 100 million people are thought to be at risk in just four countries: China, Bangladesh, Egypt and Nigeria – and that's without factoring population growth into the equation.

Sea level rise is particularly important because for various reasons – such as trade, fishing and fertile soil – humanity is overwhelmingly concentrated in coastal areas: globally a third of the world's entire population lives within a hundred kilometres of the sea,[14] and more than half of the world's twenty largest cities are located on the coast.[15] Although the most valuable real-estate in places like Manhattan or Miami is likely to be protected by sea walls for the foreseeable future, it will be impossible to enclose all the world's affected areas with concrete.

In the still longer term – over centuries to millennia – no amount of sea walls will save even the most valuable locations from disappearing under the waves. Sea level rise is such a slow process that once started it's almost impossible to reverse, and even under a moderate global warming scenario the Greenland ice sheet is likely to eventually disappear altogether. None of us will live to see this, of course, but it doesn't change the hard fact that there's enough water locked up in the Greenland ice cap to raise global sea levels by seven metres.[16] And if this happens, even New Yorkers had better move inland.

Truth be told, there isn't much – short of a miracle – that can save Tuvalu and the other coral atolls now. Because of greenhouse gas emissions during the past century, continuing sea level rise is now inevitable for decades into the future.

But longer-term outcomes, such as whether the ice sheet on Greenland melts or stays frozen, are still very much up for grabs. And it's decisions taken within our lifetimes – on whether or not to keep on burning coal, oil and gas – that will affect how the future unfolds. It's an awesome, scarcely conceivable responsibility; a choice that we make every time we start the car or switch on a light, and yet it's one that most of us don't even think about.

FAITH AND SCIENCE

I still had two days left in Tuvalu. Unfortunately, however, one was a Sunday, and on Sunday in Tuvalu there's nothing to do except go to church. So go to church I did, and sat on a pew at the front, monumentally under-dressed in my grubby trousers and unwashed shirt. Everyone else was in their Sunday best: men in the congregation wore formal grey *sulus*, and women starched white dresses and hats. Even the children had been scrubbed and dried – as the service was about to begin they all filed in looking subdued and expectant, clutching battered Tuvaluan-language Bibles.

The church was high-ceilinged and airy, with plastic flowers around the altar; a variety of wall hangings depicted scenes from the New Testament – including one of a crucified Jesus being speared gorily in the ribs. On the dot of nine the pastor appeared at the front, a bell rang once from somewhere up above, and everyone stood and launched into a lilting hymn in Tuvaluan to a tune that I found strangely familiar.

I sat next to seventy-eight-year-old Hosea Kaitu, who offered to talk to me afterwards. I gathered fairly early on that he had a different perspective on the whole sea level rise issue from everyone else I'd so far met, and – it being a Sunday after all – I considered it my duty to find out more about the Tuvaluan religion as well.

The Tuvaluan propensity to attend church isn't just a hangover from the old missionary days – it's partly also down to the genuine enthusiasm of the newly converted. Christianity first arrived in Tuvalu in 1861, when a Protestant deacon called Elekana was blown 2400 miles out of his way on a short journey between two of the Cook Islands.

Elekana ended up shipwrecked on the Tuvaluan outer island of Nukulaelae, starving and nearly dead of thirst after his eight unexpected weeks at sea. Once revived with some coconut toddy, he immediately embarked on sharing the word of the Lord with his local rescuers. After being approved by the London Missionary Society, Elekana went on to become the island's first pastor, and the spot where his boat first came ashore is today known as 'Salvation Point'[17].

Although Elekana's arrival is still celebrated in Tuvalu, subsequent missionaries sometimes let their zeal for saving the natives' souls get the better of them. Even Arthur Grimble was shocked at stories of white missionaries dancing amongst the splintered skulls of Tuvaluans' revered ancestors, laughing and shouting 'Where are the spirits of this place? Come and strike me dead if you can!' whilst the village elders who had kept the shrines sat in their houses and watched the desecration with empty faces. But the brutal tactics seem to have succeeded, and you'd be hard pressed to find a single Tuvaluan nowadays not describing themselves as a Christian.

I sat on a pandanus mat in Hosea's house, whilst the old man regaled me with stories about the war. He still remembered the day in April 1943 when Japanese planes ('like little white birds') bombarded the American air base on Funafuti, killing one Tuvaluan and half a dozen Americans. Many more were saved by a sharp-witted US airman who warned them not to shelter in the church minutes before a Japanese bomb plunged through its ceiling.

Hosea's house was full of people, including his wife, daughter, a constant stream of neighbours, and innumerable grandchildren, who played and climbed across his knees as we talked. 'You're a very cheeky boy,' he said affectionately, ruffling one small child's hair.

'I was very frightened when I heard these scientists say that all these low islands will be full of water. But I don't believe it,' he declared, brandishing his Bible at me and translating from Genesis: 'Only the Creator can flood the world. God promised Noah there would be no more flood – that's why he showed the rainbow. So every time we look at the rainbow we know the covenant that God gave to Noah.'

He chuckled good-naturedly. 'I believe in God – I don't believe in scientists.'

It's not just devout Christians who deny the evidence that Tuvalu is being hit by rising sea levels. Support has recently come from a surprising quarter: the group of Australian-based scientists responsible for monitoring the output of Funafuti's sea level gauge. Based at the National Tidal Facility (NTF) in Flinders University, Adelaide, the scientists declared in June 2001 that 'the last eight years of data return show a rate of 0.0 mm per year, i.e. no change in average sea level over the period of record'.[18]

A flurry of media interest followed. Radio Australia told its listeners that fears of sea level rise around Tuvalu 'are not supported by scientific data',[19] whilst Agence France Presse later produced a report titled 'Global warming not sinking Tuvalu'.[20] NTF's senior scientist Bill Mitchell was quoted in the article as being 'confident that the data showed Tuvalu was no more sinking than Australia was'.

The controversy was a gift to global warming sceptics. The Greening Earth Society (funded by the coal industry) carried a story on its website, as did 'Still waiting for Greenhouse', a website run by virulent anti-green John Daly before his death in 2004.[21] He used the NTF report to claim

The River Ouse bursts its banks in early November 2000, flooding parts of central York.

No pints were served at the King's Arms in York this day, during the same floods.

Wilson Sam holds up part of his hunting catch – two geese, which he shot with his grandson in the wilderness around Huslia village, Alaska.

Three generations of the Weyiouanna family, outside their house in Shishmaref. Clifford is on the left.

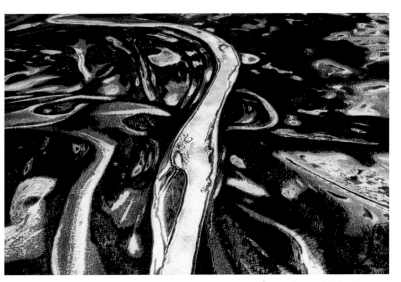

Patterns made by snow, forest and water in the interior of Alaska – this photo was taken on the flight between Huslia and Fairbanks.

Permafrost thaw damage at a house near Fairbanks. When the cabin was built, forty years ago, the land was flat!

The view from Pump Station 1, Prudhoe Bay, looking down the Trans-Alaska Pipeline as it starts its 1200-kilometre journey south to Valdez.

The graveyard in Kaktovik.

A fisherman surveys what is left of Tepuka Savilivili, the first islet on Funafuti atoll, Tuvalu, to be destroyed by the rising seas. The bigger island of Tepuka is still intact in the background.

Women carry bowls of food through the floodwaters in the centre of Funafuti atoll to a gathering in a nearby community centre. The tidal floods, brought on by rising sea levels, were up to a foot deep in places.

Toppling coconut trees on the lagoon side of Funafuti atoll, undercut by beach erosion linked with rising sea levels. This picture was taken during one of the highest tides: there's normally a narrow beach between the trees and the sea.

Waves from the lagoon wash into a family backyard in Funafuti atoll, Tuvalu, during the high tides. The shack in the middle is a chicken shed, full of very upset, bedraggled chickens.

Our Communist Party police car caught in the duststorm outside Wuwei, Gansu province, China.

The front of the duststorm racing towards Yang Pangon village in Inner Mongolia. Twenty seconds after taking this photo I was enveloped, and hardly able to breathe.

A resident of Yang Pangon, clad in the traditional blue Mao suit, shovels wind-blown sand away from his perimeter wall. In a couple of days it will all have blown back again.

that 'there is no sea level rise' and that 'the reported "plight" of the Tuvaluans is not about sea level rise at all, it's about over-population'. The Competitive Enterprise Institute, a US-based far-right think tank partially funded by Exxon, jumped on the bandwagon too, running an article sarcastically titled 'Pity poor Tuvalu – it's not sinking'.[22]

The Tuvaluans smelt a rat. They knew that NTF ran a series of state-of-the-art tidal gauges right across the Pacific region, and that the scientific data itself was likely to be sound. But why in that case was the flooding getting worse every year? And why the saltwater intrusion into the *pulaka* pits, and the accelerated erosion all around the islands? The Tuvaluans suspected that NTF, whose project was funded entirely by the Australian government, might have been coming under discreet financial pressure to produce more 'convenient' results.

No direct evidence of foul play has ever emerged. But the old adage of 'lies, damned lies and statistics' springs to my mind. Most straightforwardly, the NTF scientists had made pronouncements based on only eight years of data, although as Professor Patrick Nunn had told me in Fiji: 'You need at least thirty years of data, preferably fifty years of data, before you can detect the exact long-term signal of sea level rise.'[23]

NTF has also interpreted longer sea level records, collected by a different (and less reliable) tide gauge also based on Funafuti, and used them to suggest that sea level rise was minimal since as far back as 1978.[24] But when independent expert John Hunter re-examined the evidence, he found that NTF had got an artificially low average by taking their data only up to 1998 – a year of anomalously low sea levels due to the short-term effect of El Niño.[25] Take out the El Niño distortions and the Funafuti sea level record agrees with others around the world – and sea levels are rising, just as the Tuvaluans have long been saying.[26]

So, what is the scientific explanation for the flooding? John Hunter plotted the high tide extremes onto a graph, and found a rising trend of half a centimetre per year – five times the background rate. This is partly counterbalanced by lower low tides, explaining the less dramatic average. But as Paani had explained earlier, it is the extremes that matter: 'You can stand with one hand on a hot stove plate and the other in a deep freeze, and on average nothing is happening to you. Does that mean it doesn't hurt?'

Hunter concluded that if this increase in high extremes had been sustained for the previous fifty years it would have contributed a quarter of a metre to the highest sea levels, which would 'certainly cause an observable [i.e. to the islanders] increase in flooding during surge events'.[27]

TEPUKA SAVILIVILI

I had one last place to visit before leaving Tuvalu – the destroyed islet of Tepuka Savilivili. Former Prime Minister Toaripi Lauti had spoken about its loss with grief – for him it had summed up the urgency of dealing with global warming at an international level. Fishermen I had met by the shore remembered waking up one morning after a high wind and noticing that it had vanished.

For me, it had a special resonance. Here, it seemed, was the first Tuvaluan island to go under. A harbinger of a much greater destruction, Tepuka Savilivili gives a glimpse into the dark future that may well lie ahead for the rest of Tuvalu's atolls.

This time I wanted to cross the lagoon with a guide, and Tuvalu's head meteorologist Hilia Vavae seemed the perfect candidate.

We sped off across the lagoon, our boatman Iakopo at the back with the tiller, and Hilia and me perched on a makeshift bench at the front. The boat bounced and thwacked across the waves as Iakopo pumped the motor up to full speed. Flying fish launched themselves into the air at our side, skimming the water's surface with their elongated fins, and a lone sea turtle turned and dived into the depths as we zoomed by. On our right was the Maritime Training Institute on Amatuku, which was soon replaced by a series of uninhabited islands, all thickly forested and with pristine beaches, as we neared the far side of the atoll.

I didn't see Tepuka Savilivili until we were only about 300 yards away – it was just a low disturbance above the water, little more than rock and bleached-white rubble, with angry waves breaking on the reef flats all around. Iakopo cut the engine as patches of purple coral began to appear, and navigated expertly between them to the shallows. Hilia and I leaped out, as Iakopo stood up to steady the boat against the breaking waves. Then we waded together up onto the rocky rubble crown that was all that remained of Tepuka Savilivili.

It was only fifty yards across, and from the line of flotsam all around I could tell it would be reduced to a tenth of that at high tide. Just a half-buried pandanus log and a couple of coconut stumps broke the monotony of coral fragments, rock and coarse sand. In the blistering heat it was an unforgiving, hostile place, and I longed to be on one of the cooler, wooded islands visible just a little further along the reef. Hilia found a sprouting coconut and we scratched a hole in the rubble for it with pieces of wood, burying it so that only the green sprout showed.

'Do you think there's any chance of it growing?' I asked.

She shook her head. 'No, I don't think so. It's the other way round. The grass on other islands is already becoming brown, and will eventually disappear. The vegetation and soil will change, there will be less and less land, and finally we will get something like this island here, where nothing will grow.'

The floods on Funafuti were the beginning of this process. The high tides eat gradually away at the land, taking some of the sandy soil away with them each time they retreat. As a result, the vegetation declines, as does soil fertility. Then the waves begin to wash right over it. 'Finally it's like this.'

I asked whether she thought the floods on the main islands had become more frequent.

She nodded vigorously. 'They have increased tremendously. Take for example last year. We were flooding during the high tides in November, December, January, February and March. When I first started work in the Met

Office in 1981 this didn't happen. You would normally see it in February only. Now it's nearly half the year.'

Iakopo had brought the boat back close into the reef, and we waded out to it, climbed aboard and motored over to the nearest forested island. It was a holiday brochure view of paradise: a pristine white beach topped with graceful coconut palms, surrounded by calm blue seas, and without a soul in sight.

As I crunched up through the soft, warm sand, it seemed like the sort of place a millionaire would love to 'own', perhaps to build an exclusive mansion he could visit by helicopter once or twice a year. But here it was everyone's to enjoy, and after Iakopo had tied up the boat, Hilia and I walked along the dazzling white strand. In the shallows along the beach, two baby reef sharks – each barely a foot long – were playing in the warm water. I let out a long sigh. This was what I had been waiting for.

But Hilia had something on her mind.

'Look,' she called over. 'Look at this erosion. The last time I was here was in 1997, and it wasn't as bad as this.'

I followed her reluctantly to the high tide line. Two coconut trees had toppled over, their dying roots protecting a single spit of land. Further along, a pandanus tree had fallen, and most of the beach was taken up with a mess of cracked and tangled branches. She explained that once the tree roots were undercut, all the soil and sand could be washed away. This island was disappearing too.

*   *   *

123

We sat near the boat to watch the sunset. Even by Tuvaluan standards it was glorious – the cloud tops lit up in a dizzying array of reds and oranges, the colours reflecting off the glassy surface of the lagoon and staining the white sand pink. A warm breeze gently disturbed the tops of the coconut trees, and a single white crab skittered around in the gathering twilight. Searchlights from small fishing boats began to blink on one by one, matched only by the stars overhead.

It seemed in that moment as if the rhythms of Tuvalu were eternal – the people, the sands and the sea – all destined to stay here unchanged far into the future. But I knew that it was not to be. Instead these beautiful atolls and their warm, generous people were just fragments of a human civilisation which had flowered all too briefly, only to be cut short by the actions of others half a world away.

Sitting on our disappearing island in the middle of the calm Pacific, Tuvalu's future seemed depressingly clear – the atoll abandoned and still, its young people taking on a new life in a land far away, their homeland becoming a place of myth, a half-remembered dream, one of those magical places of longing that can never be regained.

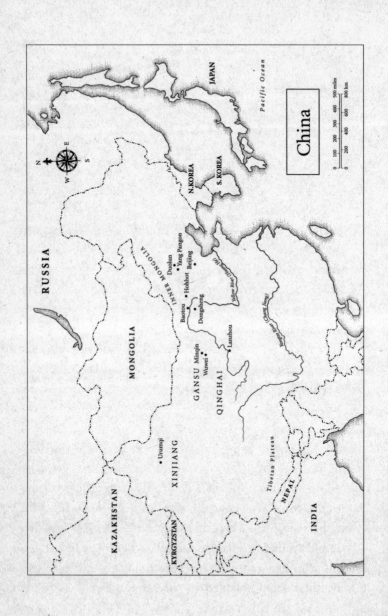

4

Red Clouds in China

Tiananmen Square was almost empty. Only one or two intrepid cyclists pedalled along its northern edge past the Forbidden City, where Chairman Mao stares down impassively from his fading revolutionary portrait. A few tourists clustered underneath to snap souvenirs, but even the city's normally heavy traffic seemed subdued.

It was spring 2002, and Beijing was suffering another of its increasingly frequent 'dust days'. Close up the dust was almost invisible, making its presence felt mainly as an irritating dryness at the back of the throat. But in the distance the whole skyline was blotted out by the dirty red-orange haze, and even at midday the sun threw only faint shadows onto the concrete pavements. Water lorries plied up and down the main thoroughfares, spraying water onto the street surface in an effort to keep down the dust, but the rising wind kept blowing swirls of it off building sites and around the deserted street corners. The few pedestrians who had ventured out strode quickly and purposefully, their faces shielded by wraparound veils or white surgical masks. I too kept my head down as the choking gusts blew down the street towards me.

The water lorries were fighting a losing battle however, because most of the dust in the air over Beijing that day hadn't come from unwashed streets – it had blown in from Inner Mongolia and the other drought-scarred provinces of North China. Every spring fierce winds lash these arid upland plains, whipping up swathes of dust hundreds of miles wide which roar down onto the Chinese lowlands – often reaching as far as Korea and Japan, and occasionally even crossing the Pacific and casting a shadow over America's Rocky Mountains.

Many scientists have linked China's spreading deserts, worsening droughts and killer duststorms to the impacts of global warming, and one of my main aims in visiting the country was to learn more about how likely this was. But I also wanted to experience the situation in the affected areas, and to speak to the people who lived there. This would mean travelling north into some of the least-visited areas of China – at the height of the duststorm season.

DUOLUN, INNER MONGOLIA

'You are very tall,' were Su Yi's first words the next day when he joined me at the bus station in a busy Beijing suburb. A researcher working with the Desertification Institute, he was in his thirties, with black hair, big glasses and clearly something of a sense of humour. 'I think you are looking very English today,' he went on, with a cheerful smile.

'Thanks,' I said, noticing that several dozen bystanders were staring quite openly at me.

He taught me Chinese characters the rattling bus negotiated the Beijing traffic, pointing out that *bei* means north, and *jing* capital, so Beijing was the ancient 'northern capital' during Imperial Chinese times, just as Nanjing was the 'southern capital'. Beijing had once been the centre of the world's greatest and most technologically-advanced empire, he reminded me, leading Europe by centuries in scholarship and invention.

But I could see that what little was left from the old days was now vanishing quickly as modern China joined the capitalist West. No longer were there great crowds of bell-ringing cyclists plying their way down the city's broad streets – instead buses, taxis, motorbikes and sleek new cars had banished the bicycles to side streets and back alleys. The traffic noise was deafening, and pollution lay in a dirty grey pall over the entire city.

The alleys themselves were disappearing too. These grimy low-rise *hutongs*, still in many ways the centre of Beijing life and culture, are lined with market stalls, street kitchens and filthy shared toilets. People live in secluded courtyards, with ramshackle houses piled on top of each other without a thought for building regulations or privacy. Every morning old women come out into the street to hawk and spit into the gutter, and beggars drag trolleys around to pick up unwanted garbage. Occasionally cars try to negotiate the labyrinthine alleyways, beeping furiously as old men hobble obstinately in front of them, stalls and pedestrians squeezed tightly in on each side.

The government sees these *hutongs* as antiquated, and almost all of them are slated for destruction – to be replaced with gleaming corporate office blocks and enormous, impenetrable banks. As we passed, I could see white paint dashes on the doors of many of the old buildings, rather like the chalk crosses marked on infected houses in medieval plague-ridden London. I couldn't read the characters, but it was obvious what this meant: 'condemned'. In some places only the facades were left; and behind the outside walls lay a rubble graveyard of old brick and ancient imperial-style ceramic tiles.

We left Beijing via a congested dual carriageway lined with poplars, heading north towards the Inner Mongolian grasslands which border the immense Gobi desert. The road led first through the Yan Shan mountains, their slopes steep and craggy, their soaring peaks lost in the distant haze. As the bus began to climb, the Great Wall itself came into view: traversing sheer mountain ridges, it was guarded intermittently by crumbling watchtowers perched precariously on each successive peak. On the rocky slopes all around fruit trees blossomed with pink and white flowers.

Further on, the terrain flattened out, the mountains suddenly giving way to farmland. Despite the hot sun people were hard at work, digging up maize roots with broad, flat spades and preparing the earth for the next crop. In smaller plots nearby, neat green squares of spring onions were well established, each bed surrounded by little ridges of soil to guide irrigation water around the plants. The narrow road was swarming with schoolchildren on bicycles, and as we neared a small town a platoon of soldiers from the People's Liberation Army jogged by in their green uniforms.

Then we were back in another range of mountains, climbing up towards a pass which led in turn onto an enormous plateau of dusty grassland – my first glimpse of the high Inner Mongolian plain. It was vast, with low brown hills massing row upon row towards the horizon, grazed only by a sprinkling of sheep and criss-crossed by gullies and eroded watercourses. Behind the occasional wall, sand had gathered in stealthy drifts.

It was well after dark when a twinkling of lights on the wide plain announced our arrival into Duolun. The Duolun County Forestry Bureau director Mr Sun Ming Shan was waiting and before taking us out to his favourite restaurant, ushered us into the only hotel allowed to accept outside visitors ('Pointed unit for foreign tourist' it declared on a polished brass plate outside). The restaurant was basic, with a concrete floor and stoves with long pipes to heat the room. Even then it was still chilly, and everyone kept their coats on.

The meal was a succession of the sort of animal off-cuts that in England people assiduously avoid or hide in things like sausages. Pig tongue was followed by backbone and sliced ear, the latter disconcertingly crunchy. Feet and stomach were also included, accompanied by a variety of rather more appetising vegetable dishes. Mr Sun ordered a bottle of strong rice wine, which we drank in little ceramic tea-cups to a chorus of *gambei* – Chinese for 'bottoms up'.

'There's a duststorm on the way very soon,' said Mr Sun with an enigmatic smile. 'Have you brought a mask? Here it can get very bad.' I noticed that the windowsills outside the restaurant had little piles of dust in the corners, the legacy of a storm two weeks before.

These North Chinese plains have been doubly hit by drought and desertification in the last few decades, and severe duststorms are on the increase. Indeed, dust hazes such as that affecting Beijing on the day I arrived would hardly even merit a mention in these northern provinces. They suffer a much worse danger – the so-called 'black wind', the strongest type of duststorms.

Black windstorms are more than a nuisance: they are killers. When one tore through the provinces of Xinjiang, Gansu and Inner Mongolia on 5 May 1993 the authorities described the ensuing disaster as 'like an earthquake'. A total of eighty-five people were left dead, with 224 injured and a further thirty-one missing.

Most victims were children, out playing in the fields and unable to get home before the surging black and red clouds engulfed and choked them. Over 100,000 farm animals were lost, whilst enormous areas of crops were simply stripped of their leaves. Visibility was so bad that people caught outside spoke later of not even being able to see their hands in front of their faces. The hurricane-force wind was so strong that its sand-blasting action even eroded away the tops of tarred roads.[1]

Although I was curious to experience a duststorm, I strongly hoped that the storm Mr Sun was predicting would not be a 'black wind'. I had no wish to be caught in the middle of this desolate plain in a deadly sandblasting tempest.

The next morning we drove out of town, arriving before noon in a small adobe village called Yang Pangon, where Mr Sun had arranged for us to spend a couple of days. The surroundings were desolate – just brown rolling hills with nothing but stones, dust and a few patches of dried-up scrub to break the monotony.

One of the first houses in the village belonged to Mr Dong, a wise-looking farmer with grey hair, brown trousers and a Chairman Mao cap, who smiled and nodded as we shook hands in the beaten-earth courtyard. Around us wandered several pigs, two cows, a puppy, lots of chickens and at least a dozen sheep, which all crowded around the water trough as Mr Dong pumped water up from a deep well.

'They get thirsty because there is no green grass to graze on the hills,' he told me, with Su Yi translating. 'All they can eat now is hay, and they're not allowed outside.'

I wandered round the village with Su Yi. It was peaceful, apart from the noise of the wind in the trees and the bleating of the animals. But outside the village it was clear that the sand was posing a major threat. All along the perimeter wall it had built up into big drifts like snow, in some places burying the stone walls altogether. On the far side of the village, further sand drifts were several feet deep against the houses, forming graceful rippled dunes in the lee of trees, walls and gates.

A man was standing on one of the larger dunes – also

wearing a blue Mao cap and traditional overalls – energetically shovelling sand away from a wall which was perilously close to being knocked down. Although the sun shone weakly, the horizon was grey – rather like on the dusty day in Beijing – and gusts of wind whipped up clouds of dust from the dry ground all across the broad, featureless valley.

Back at Mr Dong's house, a crowd of kids had arrived to gawp at the strange new visitor.

'Hello, how are you,' a bold, scruffy one shouted. 'Thenkyou very mach!' They all collapsed into hysterical giggles.

At dinner I found out the reason for all the interest – I was, apparently, the first foreigner ever to have visited Yang Pangon. 'It's true,' Su Yi confirmed when I looked sceptical. 'Mr Dong says so, and he's lived here his whole life.'

As if to emphasise the isolation, I tried and failed to find the BBC World Service on my shortwave radio that evening. There was just a cacophony of hissing, metallic grating sounds and high-pitched whistles. I lay awake that night, Su Yi snoring next to me, both of us buried under blankets on the brick platform which doubled in the daytime as an eating table. Outside the wind was still rising, and one of the cows mooed inconsolably.

We spent much of the next day touring the local tree-planting sites, being conducted round like foreign dignitaries by the village Communist Party head. Saplings were being planted in trenches about three feet apart by teams of two – many of them husbands and wives – who would take turns to open the ground with a spade whilst the other popped in a sapling and tamped down the sandy soil around it. Dispiritingly, the whole area had been planted the previous year, but because of the drought all the new saplings had died.

'He says this was once a wheatfield,' Su Yi said to me, squatting down to scuff at some brown stubble in the ground. 'But the harvest failed, partly because of the drought and partly because all the best topsoil has already blown away.' He bent the top of one of last year's saplings. The dead twig snapped off easily.

As we walked, the wind raised clouds of dust like mist in front of the nearest houses. Already my eyes felt gritty and a coating of dirt came away when I rubbed my face with my hand. It looked like the promised duststorm was on the way.

By the time we returned to Mr Dong's house, little dust whirlwinds were skittering around the yard, and the sky had turned a sickly orange. After washing the morning's grime out of my eyes with a bowl of water from the well, I ventured outside, walking by myself back to the edge of the village in the fading light. All the surrounding hills had disappeared, and Yang Pangon seemed even more isolated in the threatening grey gloom. No one else was around – except for a small boy carrying a bundle of sticks, who warned me to return by pointing at me and then at the village, before trotting quickly back towards the safety of the houses.

I hesitated, looking towards the windward direction in the northwest. The storm front was already approaching: a murky red cloud, difficult to distinguish in the fading light, but clearly heading towards me at a rapid pace. I managed to snap one photo before the first gust hit me, the sand stinging my face and neck. Suddenly dust was everywhere – in my mouth, ears, hair, eyes and lungs. Everything was red and all other noises drowned out by the colossal roaring of the wind. Coughing and hardly able to keep my eyes open, I fled back in the direction of Mr Dong's house.

Everyone was waiting on the doorstep, gesticulating at me to hurry as I opened the gate into the yard. Once I was inside the house, Mr Dong slammed the door firmly behind me and poured out a bowl of water for me to wash the sand out of my eyes. As I peered out of the window, the red sky became darker and darker, obscuring all light until it was impossible to see even a few feet across the yard. It was as if midnight had suddenly arrived in the early afternoon, eliminating all traces of the daytime in just a few minutes. The storm even penetrated the house, filling the air with a fine dust like smoke, which settled in small piles around the cracks in the windows.

Mrs Dong, who was unperturbed, swept the dust off the windowsill into a bucket and switched on the single naked electric light so that she could start serving lunch. Occasionally it cleared slightly, and the sky would be visible again through the surging red clouds sweeping across the valley. It was an odd sight – like being underwater and looking up at waves on the surface far above. A few drops of black rain spattered on the window.

Halfway through lunch there was a lull and a nearby

stand of trees was visible again, but the next brown cloud was already gathering and rushing across the valley towards us, and in less than a minute it was suddenly even darker than before. I looked at my watch. It was 2.40 in the afternoon. With the storm shut outside and Mrs Dong serving up bowls of steaming food, it felt almost cosy.

I asked Mr Dong how long these kinds of storms had been happening.

'Twenty years ago the grass around here was knee-high. These storms only started in the 1990s when the grass all dried up. Previously we would have strong winds, but no dust because the grass would hold the soil together.' Much of the problem was caused by overgrazing, he continued, but changing weather was aggravating the situation: 'This winter there was no snow at all, and the amount of rain is also decreasing. This spring was very bad – it only rained two or three times in total.'

With the storm still blowing outside, there was nothing to do except sit around talking and eating and, with repeated choruses of *gambei*, we finished off another bottle of rice wine. By late afternoon the worst of the storm had passed, and the outline of the sun was visible again through the orange sky. Thinking that morning had come early, Mr Dong's cockerels began to crow in unison.

As Mr Dong suggested, Chinese duststorms are part of a much bigger and more intractable problem – drought. I had arrived in China in the middle of the worst drought in over a century, which affected the lives of millions of rural people and eventually cost the country $1.2 billion in economic losses.[2] But the process has been silently underway for many decades: as a result of gradual climate change and rising temperatures, northern China has simply been drying up.

This creeping disaster is illustrated by the fate of the Yellow River (the *Huang He* in Mandarin), one of the world's greatest rivers, and the largest in China after the Yangtze. It runs right from the highlands of Tibet past the southern edge of the Inner Mongolian plain, and is a vital water supply for cities and crops throughout the entire region. Yet the drought, combined with rapid economic development and industrialisation, nowadays often sucks the river dry. In 1997 it failed to reach the sea at all for 226 days, with no flow along as much as 700 kilometres of the riverbed.[3] In June 2003 the government announced that the Yellow River's flow had reached its lowest level in half a century, leaving 12% of the country's entire population short of water.[4]

This drought is at the root of China's increasingly frequent and destructive duststorms. Every year, two and a half thousand square kilometres turns into desert, its sandy topsoil providing a ready source of sand and soil for the wind to pick up. Desertification is accelerating, and the rate has almost doubled since the 1950s – the nearest sand dunes are now only seventy kilometres from Beijing itself.

As a result, the number of duststorms affecting the country is increasing all the time. According to government figures, there were eight duststorms in the 1960s, fourteen in the 1980s and twenty-three in the 1990s. In the year 2000 alone, seven duststorms roared through Beijing.[5]

But the drought itself also has more direct effects, and it was to experience some of these that I wanted to travel to the western half of Inner Mongolia, where the desertification process is most extreme. I had heard reports of advancing sand dunes and disappearing lakes, but the information was sketchy at best, and very little of it has ever reached the international media. China's drought disaster is a forgotten one, taking place far away from the eyes of the outside world.

DONGSHENG, INNER MONGOLIA

Reaching the western half of Inner Mongolia meant returning to Beijing, which was fortunate because I had to meet up with my interpreter, the Beijing-based English teacher Liu Zexing. I had found his contact details completely by chance whilst surfing the internet. He was small and energetic, always impeccably dressed, and insisted that he was not the slightest bit put off by the thought of trekking through dusty deserts with his briefcase and polished shoes.

We met for the first time at the main railway station, where he joined the shouting crowds in front of one of the ticket windows and eventually emerged – already slightly dishevelled – with two 'hard sleeper' tickets for the long overnight journey to Hohhot, Inner Mongolia's industrial capital.

'Hard sleeper' was not as hard as I'd feared; there were cushions of a sort lining the bunks, as well as an itchy grey blanket and white pillows. The carriage was like an elongated dormitory, with open compartments of six beds each stacked three high along the corridor. An hour after we left Beijing, an attendant came to check our documents, and returned later with a trolley of chicken noodles in styrofoam boxes, which we washed down with green tea from a communal thermos flask. Liu then had a long conversation with his wife on his mobile phone, passing it over to me when it came to assisting with his two children's English homework. Amongst the clouds of cigarette smoke was a pleasant, comfortable hubbub, which was gradually replaced as night drew on by throaty, multi-tonal snoring.

Morning revealed a familiar landscape – our train was rumbling through desolate plains and arid rolling hills much like those I'd seen in my earlier trip to the eastern side of Inner Mongolia, five hundred kilometres away. We crossed the Yellow River on a steel bridge, the river itself far below, sluggish and dirty brown amongst an immense floodplain of pebbles and mud. Then came Hohhot city, an industrial wasteland of idle smokestacks and grim factory buildings. We spent as little time there as possible, changing to a different train and then a bus which took us to Dongsheng.

We were met by the Mongolian director of the provincial forestry bureau, Mr Alatengbao, who told us that the area had already suffered several big sandstorms that spring, arriving hard on the heels of three years of crippling drought.

'We've had drought before, but never as serious as this,'

he said over tea. And the sandstorms had been terrible: 'It was like waves turning over and coming towards you, connecting the earth and the sky.'

Mr Alatengbao was determined that I should appreciate some of the sights of the region, including the mausoleum of Genghis Khan, just a few miles out of town and the centre of a flourishing hero-worship cult devoted to the fearsome eleventh-century warlord. Even Mr Alatengbao himself – who claimed to be descended from an important clan appointed as guards to the Khan himself – personally made annual visits to the shrine to pray and sacrifice a sheep in memory of the great man.

Although Genghis Khan's rape-and-pillage exploits have made him little more than a symbol of tyranny in the West, his military successes are undeniable: at the height of their power, the Mongols controlled the largest nation the world has ever seen, an empire stretching from eastern Europe via Russia and China right down to modern-day Vietnam. Emerging without warning from the Central Asian steppes, the 'Mongol Hordes' fought with a ferocity and skill that no contemporaries could match.

But it was their skill as horsemen that gave the Mongolian conquerors the edge. Their horses were nourished by the tall grasses that grew right across the flat plains, and it is this image – of sword-wielding raiders charging through swaying waist-high grass – that endured down the ages, long after the empire itself had fallen.

As I stood looking at the countryside around the Genghis Khan mausoleum, it was impossible to believe I was in the same place. Most of the 'grassland' was now just earth, covered in places by bits of drought-resistant scrub. A few green shoots were still pushing up through the ground, but the dominant colour was brown – not a rich, earthy brown, but the dry colour of baked clay, washed over not with greenery or with water but with patches of drifting orange sand.

I later discovered that whilst staying in Duolun a few days previously I had been only a few miles from Kubla Khan's legendary Xanadu. It was equally difficult to believe that the parched landscape of that region was ever able to support a city of any size – let alone the fabled palaces and pleasure domes of Xanadu.

Not far from the mausoleum was a bleak lakebed, bleached white by the sun. Nearby stood a set of dilapidated concrete buildings in mock-Mongolian style, a tourist complex that had been specially constructed for visitors to enjoy the lakeland scenery and wildlife. Now it was abandoned, the concrete and garish paint crumbling back into the surrounding sand. Slightly further on was a small village of adobe shacks, with a dozen or so cows and sheep tucked into wooden enclosures made of sticks. A single lonely boat – the only obvious reminder that this had once been the edge of a lake – was propped up against a dead tree, forlorn against the dry, dusty horizon.

'This place is called Hong Hai Zai,' our driver was saying. 'It used to be a large area of water, but it dried up two years ago.' The scene was bare and depressing, the overcast sky and cold wind completing the sense of desolation.

In the village a middle-aged herdsman in blue overalls was drawing water from a well. Our driver dug out a packet of cigarettes, and the herdsman took two, smoking one and putting the other in his pocket. It was the first time the lake had ever dried up, he told me. The whole area had seen no significant rain since 1997.

'We used to catch fish and sell them in the market. It's hard to believe now.' He snorted contemptuously, as the cold wind blew dust from the lakebed around our feet. For as long as the water lasted the grass at the lake edge grew well and provided ample grazing for the cattle, he explained. 'Now we are getting poorer – all the cattle can have to eat these days are dried-up cornstalks.'

In the well the water level was dropping too. It had gone dry the year before and would go dry again this year unless the rains came back. 'This is the worst drought I have ever experienced. I heard from older people that there was a bad drought a long time ago, but it wasn't as serious as this.'

As if on cue, an old man shuffled towards us to join in the conversation, the herdsman's father. He also took a cigarette and told us between puffs that the lake had shrunk once before, though without drying up completely. 'That was in 1949. I've lived here all my life, and this is the worst ever. We've had no income for three years.'

He stopped to cough, hacking and spitting on the ground. Above us one of the branches of the dead trees creaked in the chilly breeze. Our driver Mr Lu lit up too, and the smoke from the three cigarettes mingled, adding to the acrid smell of animal dung and the ever-present dust.

The car got stuck in the sand as we made our way back to the road. My interpreter Liu and I got out to push, straining against the back bumper. Liu's blazer was looking rather dirty by the time we had freed the vehicle. 'It's OK,' he said with a grin, brushing himself down and getting back inside.

We drove south for another thirty miles, nearly as far as the border with Shaanxi province. There were increasingly large sand dunes piling up on both sides of the road, sometimes spilling onto the tarmac itself – tangible evidence of advancing desert. Barriers of poplar and willow trees had been planted as protection, but the dunes were pressing on regardless, in some places burying the trees right up to their crowns.

In one place a few acres of green pastureland still survived, but were hemmed in on three sides by the advancing dunes. As I walked around it, the contrast between the bright green grass and orange sand reminded me strangely of a golf course. Once the sand moved over it, this grassland too would be gone, adding a few more acres to the new Inner Mongolian desert.

Across the road two men in blue jackets were ploughing the dry soil, one leading a donkey whilst the other steered the steel plough behind. They spoke about the increasing sandstorms, and the struggle to survive without rain, their crops left stunted by a shortage of water and the ever-present desiccating wind.

I asked if anyone had lost land because of the advancing sand dunes.

They both nodded. 'Yes, we have,' one answered, pointing behind him. 'The sand dune has advanced so far that we can't plant anything. We have lost about ten *mu* [over half a hectare] and a lot of income because of the sand.'

'Was it as bad as this thirty years ago?'

'Back then there was plenty of rain,' answered his brother, leaning on the donkey, which looked like it was glad of the break. 'It's always been sandy, but because of the drought now it's easier for the wind to move the sand. This drought is the worst anyone can remember.'

So what were they going to do?

One of them shrugged. 'The situation is bad, but we'll get by somehow.' His brother nodded stoically, patted the donkey, and they set off once again to continue ploughing what was left of their parched field.

There is intense debate amongst the Chinese scientific community about the precise causes of the drought; the picture complicated by increasing human pressure on the land – overgrazing and deforestation have both contributed to the current crisis.[6] But two things are certain: temperatures are rising and rainfall totals dropping. Climate change may only have tipped the balance, but large areas of northern China are becoming virtually uninhabitable.

For decades climate scientists have repeated the mantra that global warming means more droughts as well as more floods. There is increasing evidence from other drought-stricken parts of the world that it is beginning to come true.

One crucial factor is heat. As continental interiors begin to heat up, more water is evaporated by plants and from the land surface – and drought results.[7] So higher temperatures can cause drought even where rainfall totals remain unchanged. One recent study linked this effect with the severe 2002 drought in Australia, where low rainfall was exacerbated by soaring temperatures and rapid evaporation.[8] This warming is likely to accelerate in the future, and one computer model projection shows the continental interiors of North America, Europe and Central Asia up to a third drier by the middle of the twenty-first century.[9]

The process may have already begun in northern China. The 1990s were warmer than any other decade in the previous six centuries,[10] and temperatures since the 1950s have shot up up by 1.5°C.[11] These high temperatures have been drying out the soils, helping turn vast areas into new desert and driving increasing numbers of people off the land.

And in future it could get even worse: computer models project increases in rainfall of up to a fifth by 2050, but much greater rises in temperature. The extra warming would be more than enough to counteract the smaller rise in rainfall, and northern China would get even drier still.[12]

The other crucial factor causing drought is the atmosphere itself. The processes which cause drought are the opposite of those which cause rain: rainfall is generated where air rises, condensing water into clouds and generating precipitation, whereas sinking air prevents cloud formation and stops rainfall. Global warming means the intensification of this hydrological cycle, with heavier rain in some places (and at some times) balanced by reductions elsewhere.

One study has already identified increasing trends in severe wet and dry areas around the world.[13] In addition, strong evidence has emerged recently which links a major drought affecting everywhere from the southern US to central and southwest Asia between 1998 and 2002[14] to rising temperatures in the Indian Ocean. The mechanism is straightforward: warm seas triggered heavier rain in tropical regions, with sinking air causing drought in the mid-latitudes. And Central Asia – including Tajikistan, Uzbekistan and Turkmenistan – was one of the worst-affected regions of all.[15]

Rainfall trends in northern China have indeed been declining. There have been measured reductions in both the number of rain days and the overall rainfall totals over the past half-century,[16] and conditions in the region now are even worse than in the very dry 1940s. Out of twelve recorded 'drought disasters' between 1949 and 1995, half occurred between 1986 and 1994.[17] And the 'drought disaster' that I was experiencing would turn out to be the worst of all.

WUWEI, GANSU PROVINCE

Wuwei is one of the remotest cities in China. Stuck in a narrow corridor between high mountains and searing deserts, this ramshackle city of one million sees no tourists and hardly any foreigners. The main road out of Wuwei leads eventually to Urumqi, capital of the restricted

Xinjiang province, where it splits in two – the southern section leading to Pakistan's Karakoram highway and the northern half finding its way to the border with Kazakhstan.

In the past, its location as an oasis on the old Silk Road made Wuwei an important stopping point for trading caravans, and Marco Polo himself would have passed through there on his thirteenth-century voyage into the Middle Kingdom. Now it's a neglected backwater, hemmed in on all sides by advancing deserts, and with a steadily diminishing water supply from rapidly-disappearing rivers and lakes.

I had already seen graphic evidence of this on the journey there from Lanzhou, Gansu's capital, three hundred kilometres south of Wuwei. We had left the city by car, accompanied by Dr Zhang, director of the regional water bureau, who punctuated the journey with descriptions of how the drought was affecting his region.

'You see all the hills there?' said Dr Zhang as we descended on the far side of a pass, where rows of peasants ploughed diligently between patches of thawing snow. 'They used to be green with grass. Now they're baked brown, as if they've gone bald.'

The air in the distance was already much more hazy – this was the Hexi Corridor, one of the most sandstorm-battered areas in China. It is also a closed basin: unlike those draining south into the Yellow River, rivers in the Hexi Corridor are destined never to reach the sea, and eventually peter out into huge salt flats in the desert.

Today many of the rivers don't even make it that far. Dr

Zhang stopped the car just in front of a bridge not far from Wuwei, and we all got stiffly out onto the side of an embankment. It was a dried-up riverbed, with not even a trickle of water left – just a broad expanse of gravel and rounded pebbles, which in some areas had even been mined for roadstone, making the course of the old river barely distinguishable.

'The water here used to be a hundred metres wide,' Dr Zhang announced. 'The river only runs for about ten days a year now. It can go on a bit longer if there's a flood, but the longest recently was when it ran for a whole month back in 1996. There used to be six rivers in the Wuwei area but now all of them are dried out.'

I asked him what had taken the water away.

'We have a saying, that here in this region nine out of every ten years brings drought. Now it's ten years out of ten years. It was especially bad between 1996 and 2001 – there were six consecutive years of drought.' The situation had been worsened further by the construction of reservoirs in the upper reaches of the rivers and a rapid expansion of the population of the region's cities. In the 1950s Minqin, a smaller oasis town east of Wuwei, used to receive 0.5 billion cubic metres of water a year. Now it got only a fifth of that – the rest has been diverted to other cities higher up or has disappeared altogether.

Another aspect of global warming is also affecting the rivers. The western edge of Gansu province borders with the Tibetan Plateau, and the Qilian Mountains which form the spine of this border are the source of all the crucial oasis rivers. But the glaciers in these mountains, which keep the rivers running all year round, are fast disappearing. Three-quarters are known to be in full retreat, and half of all the glacier ice has disappeared in the last 150 years.[18]

My main objective in Wuwei was to track down Professor Liu Xinmin, the sixty-year-old former director of Lanzhou's Desertification Research Institute, who knew the area intimately. I hoped he could tell me more about how rapidly the deserts were encroaching and why. I had already met him briefly in Beijing, but he had failed to show up in Lanzhou, and was only rumoured to be in Wuwei.

But news of our arrival – a visiting foreigner is unusual in these parts – must have travelled fast. We were just unpacking in a small hotel room when Professor Liu unexpectedly swept in, looking very dapper in his pork pie hat and corduroy jacket.

Like many Chinese people his age, Professor Liu has been through a lot. The old Chinese curse: 'May you live in interesting times' struck his generation perhaps more than any other. Banished from university to hard labour on a military farm during the Cultural Revolution at twenty-six, he nevertheless managed to secretly learn English. The only reading matter officially allowed was Chairman Mao's *Little Red Book* of quotations, which he got to know entirely by heart. And although no one memorises the *Little Red Book* any more, Professor Liu admitted: 'Mao said one thing that I still believe: "Serve the people". You should always serve the people I think.'

This commitment has formed the basis of Professor Liu's

recent work, which focuses on studying the impact of desertification, and – through various schemes such as tree planting and dune stabilisation – trying to mitigate it. With groundwater disappearing so fast, the tree planting has been a mixed success, he told me, although his latest project to stop sand dunes moving onto farmland by pouring asphalt all over them seemed to have potential.

He took us that evening to one of the sites nearby, where a rather shapely sand dune had been encased entirely in black pitch, which had been poured into hundreds of adjacent squares, each bordered with straw. The idea, he told me, was that any rain that fell would soak through the pitch, but would then be prevented from evaporating again, and would therefore allow any plants to get a firm roothold. He assured me that the worst pollutants had been removed from the pitch, and that it would biodegrade in five years, but I wasn't convinced. It really *didn't* look very nice either, certainly when compared to the graceful reddish curves of the surrounding dunes – some of which were twenty to thirty metres high.

But as the crisis is becoming critical, aesthetics are very much a secondary concern. Not only are rivers drying up, but entire lakes have disappeared, and villages have been emptying of their people – leaving whole areas of formerly productive farm and grazing land abandoned. In addition, ever more regular sandstorms are sweeping the entire area, and with each storm the surrounding deserts creep a little closer.

*   *   *

The following morning we were all standing around on the hotel steps, which fronted onto a potholed and rubbish-strewn side street, when our vehicle arrived – a four-wheel-drive black-and-white jeep with a red light on the roof. Professor Liu was an influential man, and the local Party had generously allowed us to use an official vehicle for our trip out into the desert.

'But that's a police car,' I objected. 'Surely we can't travel around in a police car!'

'Don't worry, it's all arranged,' Dr Zhang assured me, grinning. 'There are no tolls to pay in this, either.'

We all piled in and began the journey out of Wuwei, occasionally sounding the siren to help overtake bothersome slow trucks. Small tractors plied their way up and down the road, piled with such enormous loads of hay that their drivers were almost completely buried in the middle. Long greenhouses lined the fields, which bustled with people weeding, planting and watering the crops. Yields in these oasis areas are some of the highest in China, Professor Liu informed me, which is why it is so disastrous that the water supplies are gradually running out.

Much of the route to Minqin was blocked by roadworks, and we had to bump along in the desert, lurching from side to side in the rutted tracks. A few stunted bushes and pale clumps of dry grass were the only living vegetation. Everywhere else, in any of the areas no longer irrigated, the soil seemed to have given way to sand.

We stopped briefly at an enormous reservoir, now the end of what remains of the Shiyang River. Again, I could see how the wider effects of drought had been worsened by misguided human intervention – the construction of this artificial lake had sounded the death knell for the natural lake further downstream and the wetlands that used to surround it, and it is there that the desertification problem is worst.

Minqin region is currently among the most badly desertified areas in the whole of China. In ancient times it attracted government attention for its 'abundant waters', 'fertile soil' and 'green pasture'. Now 94% of the country territory is desertifying, and sand dunes from the surrounding deserts are encroaching at an average of five to ten metres per year.[19]

The deserts are nothing new, Dr Zhang told me. What has changed is that they are now expanding so rapidly. Abruptly, he pointed out of the window at the big sand dunes on the left side of the road and then indicated towards some more dunes in the distance on the right. I could see them shimmering through the heat haze, squeezing the narrow green strip along the road in their relentless onward march. This was the spot where two of the biggest deserts are joining up, in a deadly pincer movement which is condemning the remaining oasis land around Minqin to gradually wither. Only about a kilometre of green land still separates them – and it too will be overcome within the next few years. Then Minqin and its oases will be cut off, an isolated remnant in a spreading tide of sand.

We stopped in the town for lunch and tea, each cup served with an enormous lump of rock sugar poking out of the top. Professor Liu ordered pigeons for us all – small, skinny birds which arrived plucked and gutted, but apart from that had simply been boiled whole. Dr Zhang demonstrated the correct way to eat them, delicately nibbling the skin off the head of his pigeon, then cracking open the skull like a nut and sucking out the brains, but I couldn't bring myself to follow suit.

Back outside I noticed that the wind was rising steadily, and that the sky had the same ominous yellow-brown tinge that I had last seen before the storm in Duolun, Inner Mongolia. Dr Zhang had also noticed the change. 'Maybe it isn't the best idea to visit the lakebed when a duststorm could be coming. Are you sure you want to go?'

'Well, it's my only chance.' Professor Liu had told me that a village on the edge of the lake had been completely abandoned, and I was eager to see what the area looked like.

Back in the desert, sand was blowing across the road in winding orange streaks. On the left-hand side gnarled and stunted trees struggled to hold their own against the shifting sand.

'Those are Russian olive trees,' Professor Liu identified. 'They were planted here twenty years ago, but the water table has fallen so much that they can no longer survive.' Some of the trees were already dead, their trunks half-buried in dry dunes. Others had just a few bits of foliage left on a couple of the higher branches.

The lakebed itself was even more desolate. Only a few small bushes offered any protection against the wind, and clouds of dust were already rising, whipped up from the flat expanse.

'This is Qingtu Lake,' shouted Professor Liu over the wind. 'We're in the centre now. In the 1950s it was water here, then in the 1960s and '70s it was wetland. Now it's completely dry. But if you dig here you can sometimes find old watersnails.' My eyes were already stinging again with the dust and I had to return to the car to make notes, but Professor Liu seemed completely unaffected, kneeling down to scratch a hole in the ground, searching determinedly for the long-dead snails.

On the near side of the former Qingtu Lake was the abandoned village – which according to Professor Liu had once been prosperous before the deserts moved in. I had a look around the ruins. Bits of straw and twigs were blowing around in circles in the remains of a front yard. The roofs had all collapsed, but some of the holes for washbasins remained, and in one house an old fireplace oven still stood. All around, dead and dying trees creaked in the gathering wind.

Despite all the evidence, it was difficult to believe that the community of several hundred people who once lived here had been turned into environmental refugees without anyone noticing. But not a single journalist had visited, and all anyone knew was that the former inhabitants had gone to join relatives or to eke out a living doing manual labour on the edge of the cities.

It was the same story in many villages throughout the area, Professor Liu told me. Indeed, I had seen the same thing over a thousand kilometres away in eastern Inner Mongolia. Not far from Yang Pangon, a cluster of ruined walls were the only remains of a village once home to five hundred people. No international compensation came to all these displaced people: once their farmland dried up, they had little choice: migrate elsewhere or starve.

But not everyone has gone from the desiccated edge of Qingtu Lake. Across the rough track one house remained inhabited. We knocked and hooted for some time before the wooden courtyard door was reluctantly opened by a middle-aged woman with two silver teeth.

'I thought you were the police!' she cried in a relieved voice after Professor Liu explained the reason for our visit. Suddenly friendly, she invited us in. Chickens pecked around the neatly-swept quadrangle, and various barn doors swung and banged in the wind. On the far side of the courtyard was the house; a small, single-roomed building with a bed, cooker and hard sofa. Old Communist Party propaganda posters were plastered around the whitewashed walls inside, but apart from that the room was bare.

The woman, Ye Yinxin, had lived in the house ever since moving in with her new husband thirty years ago when the areas around were still productive wetlands. Everyone else had cleared out of the village in 1998. She was the only one left, the single remaining inhabitant of this crumbling adobe ghost town. Even her husband had recently left, searching desperately for manual jobs in Inner Mongolia to bring in some cash.

We all arranged ourselves around the room: Professor

Liu, Dr Zhang and my interpreter perched on the edge of the bed, leaving me – as the honoured guest – in the single chair, which was placed strategically in the middle of the floor. Mrs Ye stood, her calloused hands firmly at her side.

Rather tactlessly, I began by asking her whether she was lonely.

'Of course I'm lonely!' she answered fiercely. 'Can you imagine how boring this life is? I can't move, I can do nothing. I have no relatives, no friends and no money.'

I asked what life was like before the area turned into a dustbowl.

'I used to live a good life,' Mrs Ye replied, more softly. 'There were lots of people around and we all grew crops. People used to help each other out a lot, and we always had time to visit each other in the evenings.' She told me how the weather had changed too: in the old days it had rained a lot, and water was plentiful. Now there was continual drought, and when rain did fall no plants were left to soak it up, so it just evaporated away. She still kept a few sheep and cows, animals which were able to drink the saline water out of the well and eat brought-in hay. But for clean drinking water she now had to travel twelve kilometres in a tiny tractor to fill up a barrel.

'How long will you stay here?'

She laughed bitterly. 'Come back in five or ten years and I'll still be here. There's nowhere for me to go.' Her parents lived about thirty kilometres away, but life there was only slightly better, and they had no spare land. Her best hope was that her husband would get a good job and return with some money. But she hadn't seen him now for three months, and he had no way of sending word. Even her children had abandoned the home – both now lived in Minqin town, where the youngest was still at school, and neither of them had the money to visit.

I looked over at Professor Liu. Normally impassive, he was visibly moved. He wrote something down on a scrap of paper for Mrs Ye, and handed it to her quietly as we left. 'This is hard living,' he muttered, as we filed one by one out through the low door. 'Very hard living indeed.'

Professor Liu just shrugged when I asked him what he had written on the paper, but my interpreter told me afterwards that he had left his contact details with Mrs Ye, and would help in any way he could, perhaps assisting her children to apply to university. I was touched by his generosity, especially as I had just sat there feeling useless: there seemed like nothing I could do to help, and she had refused payment for the interview. She had even apologised for not having any fresh water to offer us, explaining – as if it was somehow her fault – that the water barrel in the yard was empty.

Back on the other side of Minqin town more sand was blowing across the road, and I could tell that the storm would not hold off much longer. Rows of poplars surrounding the nearby fields were thrashing around in the strengthening wind. Dr Zhang peered out of the side window, looking concerned. Above us a bank of cloud was rolling in, blotting out the sun. It looked like a storm front, purplish higher up but changing to red closer to the ground. We watched it move closer, the boiling red mass chasing across the fields towards us.

Everyone rolled up their windows and seconds later it slammed into the car, enveloping everything in a choking red mist. Visibility was suddenly down to less than fifty metres and the wind was roaring all around. Other vehicles on the road switched their lights on and slowed down. In the fields people were packing up urgently. One old man had a coat wrapped around his head, and was struggling to get home against the strong wind. The workers from one of the road crews we'd passed earlier were crouched behind a stone wall, their shirts held up against their faces.

'This is the fifth duststorm in the region so far this year,' said Dr Zhang sadly. Our driver switched on the flashing police lights, which lit up the dust all around us like a strobe light in a smoky nightclub. At one point I got out to take a photo, coughing and choking, my camera barely functioning and dust getting in my eyes, nose, mouth and hair.

As we drove past the big reservoir we'd visited earlier there was a lull; but on the other side the wind rose quickly back to gale force, reducing visibility to only about twenty metres. Sand was blowing in front with such force that in places the road itself was nearly invisible. There was a hissing sound as it blasted itself against the right-hand side of our vehicle.

As we headed back to Wuwei, the wind roared louder still, killing any conversation, and we all watched silently as it blew the brown soils of China up into the air and away.

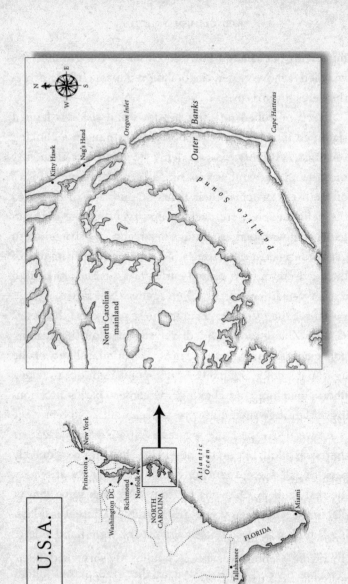

5

Hurricane USA

It began as a normal Monday morning. I ate a leisurely breakfast, then took a cup of coffee upstairs and sat down to check my email. The weather was dull, and a few drops of September rain were beginning to break the summer-long drought. Outside parents were walking their children to class, and shouts from the school playground drifted up my road. Yawning, I made my usual check on the website of the US National Hurricane Center.

So far it had been an unremarkable year – not a single hurricane had yet formed in the Atlantic, and we were already halfway through the season. A couple of weaker storms had come and gone, one called Edouard dropping lots of rain in Florida whilst another called Dolly travelled harmlessly north into the mid-Atlantic. But my Monday routine was about to be interrupted – on the National Hurricane Center website a 'public advisory' had been posted about a new storm. Its name was Gustav.

'GUSTAV THREATENING MID-ATLANTIC COAST . . .' the NHC was warning, in its customary dramatic capitals like an old telegram. 'A TROPICAL STORM WARNING IS IN EFFECT FROM CAPE FEAR NORTH CAROLINA NORTHWARD . . . GUSTAV IS MOVING TO THE NORTHWEST NEAR 14MPH . . . THE CENTER WILL APPROACH THE MID-ATLANTIC U.S. COAST LATE TONIGHT OR EARLY TUESDAY. MAXIMUM SUSTAINED WINDS ARE NEAR 45MPH [seventy-two kph] . . . WITH HIGHER GUSTS. SOME STRENGTHENING IS FORECAST DURING THE NEXT 24 HOURS.'

I knew that I had to get a close-up view of how tropical storms worked: and the best way to do this would be to experience one at first hand. Now I had less than a day to get to North Carolina before Gustav hit. I clicked offline and picked up the telephone to call a last-minute bookings service.

NORTH CAROLINA

I was out of the house less than two hours later, having made desperate last-minute arrangements for someone to feed the cat. During the bus journey to the airport I used my mobile phone to cancel everything I was supposed to do in the following two weeks. As the bus left the motorway and headed for Heathrow's Terminal 4, I tried to get to grips with the reality of the situation. This was no longer the start of a normal week. Now I was flying to America to intercept a strengthening tropical storm.

Like most US tropical storms, Gustav had begun as a swirling mass of thunderclouds far out in the Atlantic. Nourished by warm sea temperatures, the swirl had gained energy and begun to form a vortex, taking moisture high up into the atmosphere where it condensed to form more clouds and rain. This process released the heat that was initially used to evaporate water from the ocean surface, driving more upwards convection and intensifying the system into a tropical storm. The strongest tropical storms become hurricanes:* the most destructive large-scale weather systems found anywhere on Earth.

Unlike tornadoes, which are narrow funnel clouds measuring only a few metres across, hurricanes can be gigantic – often several hundred kilometres in diameter. In the Atlantic classification system a hurricane begins life as a 'tropical depression', which then strengthens into a 'tropical storm' and is only declared a 'hurricane' once the winds reach 120 kilometres per hour (seventy-five miles per hour). As I crossed the Atlantic, heading for Washington DC, Gustav was already strong enough to be called a tropical storm – but how long it would take to become a hurricane was still open to question.

The plane landed at Washington Dulles Airport at 10 p.m. local time. I hired a car and spent the first half hour on the road in a state of near-panic, trying to figure out what all the signs meant, as well as how to drive an automatic car on the right-hand side of the road. By midnight I was tired as well as jet-lagged, and for safety's sake checked into a drab Holiday Inn somewhere to the north of Richmond, Virginia.

---

* 'Hurricane' is a US term. The same storms are known as 'typhoons' in the western Pacific and 'cyclones' in the Indian Ocean.

I switched the TV in my room over to the Weather Channel, which was giving regular reports on Gustav's progress. It was still far enough away to allow a few hours' sleep, but when I woke at 5.30 the satellite picture showed an impressive revolving swirl of cloud heading straight for Cape Hatteras, on North Carolina's narrow Outer Banks. The Weather Channel's reporter was doing live dispatches from the rough-looking seashore, informing viewers that schools in the area were now closed, as were tourist facilities – but no evacuation of residents had yet been ordered.

Back on Interstate-95 I continued south, soon turning southeast towards Norfolk, Virginia, through a heavily industrial area of shipyards and docks. The weather was still fine, but when I crossed into North Carolina I saw the edge of the storm for the first time. It was a long line of high cloud in the sky, almost dead straight, but thicker and inky black as it dropped towards the horizon. Behind me the heavens were still blue and clear, but as I headed towards the pall of heavy cloud the sun began to fade away and the first showers of squally rain blew across the forests and marshes.

I passed through several small towns, with white-painted wooden clapboard houses and neatly-kept grass verges into which were stuck placards: 'Keep Susan Johnson Sheriff' and 'Denton Sneider for County Commissioner'. I reached the Outer Banks – a long, narrow line of barrier islands running offshore for more than a hundred and fifty kilometres – by eleven in the morning, already in heavy rain.

The local radio station was warning about storm surge flooding during the upcoming high tide. Caused by the high winds and low pressure of a storm system piling up water against the coast, these surges were predicted to lead to tides two metres higher than normal.

Although the road was sheltered by high sand dunes on the seaward side, fierce gusts of wind brought heavy rain lashing across the windscreen in strong bursts which reduced visibility to less than fifty metres. Tall wooden houses built in gaps in the dunes, normally beach-side holiday homes, were currently bearing the brunt of the weather. As I neared the built-up area, I noticed that the road in front was suddenly running with foamy water. It was flowing over the dunes, around the first house and down its drive onto the main road. The storm surge had begun.

I got out and splashed through to take a closer look, the water rushing around my ankles as I walked up the shallow slope of the drive towards the house, beyond which lay the roaring Atlantic. Each time a wave broke, another torrent of foamy water joined the river cascading past the house and down the drive. The road itself began to look in danger of being cut off – preventing any retreat if the storm got worse. Already soaked, I got back in the car but decided that there was little option but to continue. After a brief lull the weather closed in again, with such heavy rain that I had to slow right down for fear of driving off the road without realising it.

On the other side of a bridge, at a gap between the islands called Oregon Inlet, the flood tide was surging around the concrete legs of the road bridge as if in a fast-running river, choppy waves breaking against them in the rising wind. The only other vehicle in the car park was a satellite truck belonging to Fox News, its solitary camera-man filming my unsteady wind-blown progress back from the shore. I asked him on the off-chance whether he'd seen anyone from the Hurricane Intercept Research Team – a group of storm-chasing professionals I'd contacted from the UK and was hoping to meet up with. I knew that if there was a storm in the area, HIRT were likely to be close by.

I wasn't wrong. 'Yeah, they were here about a quarter of an hour ago,' the cameraman replied. 'I think they're staying at the Comfort Inn or someplace down at Cape Hatteras.'

This seemed like good luck, so I continued driving south towards the cape – the centre of the tropical storm warning area.

After a couple more heavy squalls, an uneasy lull had settled by the time I got to Cape Hatteras, and I found the Comfort Inn straightaway. In the car park was a large green Rodeo SUV with 'Hurricane Intercept Research Team' inscribed on the side and rear. On its roof was an impressive-looking array of weather equipment, including an anemometer (a windmill which measures windspeeds), a temperature and rainfall gauge, and at least five aerials. 'Be Hurricane Prepared!' advised a sign attached to the front bumper, just above the 'HIRT-ONE' number plate.

HIRT's director Mark Sudduth was pacing up and down in the shelter of an overhead balcony, talking into a cellphone. 'I can confirm that we just recorded a gust of fifty-five miles per hour [eighty-eight kph],' he was saying to a local radio station. Young and solidly built with short brown hair, he seemed motivated nearly to the point of hyperactivity. His colleague Chuck Ripple was older and a little more laid-back, a ham radio enthusiast with a similar number of aerials atop his own even bigger white SUV. They both sported identical T-shirts, prominently emblazoned with the team's logo.

HIRT's mission is to develop a better understanding of exactly how hurricanes affect infrastructure and human populations, in order to help protect against future storms. Although they're not 'hurricane chasers' in the usual sense of evacuation-disobeying recreational thrill-seekers, this still means their getting to the centre of the strongest storms they can find to take measurements – the perfect people for me to link up with as Tropical Storm Gustav bore down on the coast.

Mark Sudduth finished his call and introduced himself, inviting me into the team's temporary office – a messy hotel room which opened directly onto the car park. Spread over the twin beds were a variety of laptops, water-proofed video cameras, mobile phones and many metres of cable. We crowded round a television to watch the latest radar pictures of Tropical Storm Gustav, bearing down on the Outer Banks and still strengthening. The latest advisory from the National Hurricane Center in Miami forecast that the storm would track north and then northeast back out to sea.

'But it's going west!' yelled Mark, pointing at the screen, where a prominent loop of circulation seemed about to make its way onshore.

Mark was also trying to get his laptop to connect to the internet through his cellphone – which would give access to near real-time radar and satellite shots. Suddenly there was a click and all the power went out.

'Oh, that's good,' he complained, dropping a box of doughnuts.

However, oddly, the loss of power seemed to improve his internet connection. 'I'm through!' he shouted, as the screen slowly loaded up a colourful radar image of Tropical Storm Gustav. There was a band of heavy rain moving to the north of where we were currently based, but the storm was clearly moving northeast, just as the hurricane forecasters in Miami had predicted.

'Look at that loop!' Mark admiringly pointed at Gustav's centre of circulation, which was still offshore. 'If we go north now, we can beat that rain band,' he added. 'It's headed right for Oregon Inlet. Let's go!'

There was frantic packing up of all the equipment, and in less than ten minutes we were on the road, my small hired car sandwiched between the two official-looking flashing lights of the HIRT vehicles. The rain closed in again, accompanied by more high winds and blowing spray which at times almost blotted out Mark's vehicle just a few yards in front. With the land and sky nearly indistinguishable, only the flashing HIRT orange light was visible between the frenetic to-ing and fro-ing of my windscreen wipers.

Suddenly Mark pulled over, calling out to Chuck. 'It's the anemometer!' The blade had sheared off and the vital nut which held the assembly together had also disappeared. 'The show could be over right here,' he lamented: without the ability to measure windspeed, the team would have no way to judge the strength of the unfolding storm. Although a new propeller was quickly found, a quarter-inch nut was nowhere to be seen. Standing with my back to the wind, I held open the lid of Chuck's metal toolbox, mounted behind the driver's cab on his vehicle, whilst he rooted around for the vital spare part.

'Here's one,' he called triumphantly after a long five minutes, passing it up to Mark. It was fixed quickly, and we moved off again at a slow crawl.

A few miles to the north, where the sand dunes dividing us from the pounding ocean were closest to the road, Mark pulled over again and began filming. It was a startling sight. Streamers of sand were blowing off the tops of the dunes into the air, like powder snow off an Arctic snowdrift. The sand was blasting noisily into the side of my car. It had built up under the windscreen wipers so much that they had almost stopped working.

I got out to try and wipe it off from the outside, but my eyes were immediately shot full of wet sand grains. Straightaway both my contact lenses blinked out, and the whole scene dissolved into a grey blur. I quickly retreated to the car, scrabbling around in my rucksack for cleaning fluid so I could replace the lenses before the whole convoy moved off again. Using the rear-view mirror, I managed to get them back in just as Mark was disappearing up the road. I followed him quickly, careless of the floods across the road or the sheets of spray and sand that were blowing between us.

We all pulled over again in the car park next to Oregon Inlet. Mark had managed to get a connection to the internet up and running in his car, using a cellphone and laptop which both stayed charged up from a power connection running straight from the engine's alternator.

'We gotta stay right here!' he yelled at me whilst I peered through the window. 'Look at this radar!' The latest picture – updated every six minutes – showed an angry orange swirl still to the south of us, with a band of heavy rain just hitting our area.

This was soon torrential and blowing almost horizontally across the car park, accompanied by sheets of white spray, pounding into my back. I retreated dripping to my car which felt less than safe as the rain slammed into it and the wind rocked the whole vehicle from side to side. Mark appeared outside, wearing goggles, and fiddled around with the anemometer again, the aerials bending at crazy angles around him in the wind.

When the rain paused for a few minutes, I joined him outside and peered up at the sky. Above us I knew there was a gigantic swirl of cloud, its curvature almost visible as I scanned from one dark grey horizon to the other. The wind direction was subtly changing, backing round from the east to the northeast as the storm's centre passed by us offshore.

Then the rain came on again, even heavier than before,

and we both retreated into Mark's vehicle, me trying not to drip rainwater on his laptop. His air conditioning made it suddenly very cold. A digital display showed the windspeed as measured by the rooftop anemometer – generally hovering around forty–fifty mph (sixty-four to eighty kph) but much higher in gusts. As I watched, one hit sixty-six mph (105 kph), just ten mph short of hurricane strength. The barometer showed a steadily falling air pressure as the storm crept closer: 992.2 millibars, 992.1, 992.1 again, then down to 989.9 for the first time. Mark loaded another radar picture onto his laptop.

'There it is – the eyewall,' he said, a note of respect creeping into his voice. A tightly-wound orange circle with a hole in the middle like a doughnut was clearly visible, passing southeast of us, with a diameter about fifty kilometres across. Whilst in a fully-developed hurricane the eye itself is an oasis of calm, the wall of the eye is the most violent part of the storm, a narrow area of screaming wind and bucketing rain that can cause immense damage. When the eyewall of Hurricane Mitch crossed the island of Guanaja in October 1998 it defoliated the entire place – tearing leaves, branches and even bark off trees, and leaving a devastated moonscape behind. Gustav was not strong enough yet to cause that kind of damage, but the image we had seen in the radar showed it was developing that potential.

Mark was busy tapping away on the computer to update his own hurricanetrack.com website, posting a note at the top of the front page: 'WINDS HAVE REALLY PICKED UP, 50 MPH SUSTAINED, AND THE RAIN IS BEYOND BLINDING . . . MORE TO

COME LATER . . .' Five o'clock brought another public advisory from the National Hurricane Center in Miami. 'CENTER OF GUSTAV PASSING JUST EAST OF CAPE HATTERAS . . .' it was headlined. 'TROPICAL STORM FORCE WINDS EXTEND OUTWARDS UP TO 200 MILES [320 kilometres] FROM THE CENTER. HIGH SURF AND DANGEROUS RIP CURRENTS ARE EXPECTED ALONG THE U.S. EAST COAST FROM NEW YORK SOUTHWARD TO THE NORTHERN FLORIDA COAST TONIGHT.'

As we read, Mark loaded a satellite picture of the whole area. It was breathtakingly beautiful – revealing a gigantic swirl of white cloud, perhaps over 500 kilometres across, of almost perfect symmetry. Gustav trailed spiral arms of cloud like a galaxy, its spinning tropical moisture extending over five states – from Maryland and Delaware to the north, Virginia to the northwest, and both North and South Carolina – as well as far out across the Atlantic.

The storm was now at its height. The bridge over Oregon Inlet was almost blotted out by spray, driven off the sea by the howling force-11 wind. We moved onto it in slow convoy, halting at the top where Mark's anemometer recorded a peak gust of sixty-eight mph (108 kph), only a little below hurricane strength.

But Gustav was already slackening as we continued on to the small town of Nag's Head and sheltered on a gas station forecourt. Everyone was tired, soaking wet and getting cold, and as we watched, the changing wind direction confirmed that – exactly as the forecasters in Miami had predicted – Tropical Storm Gustav was beginning to make the slow turn which would take it back out to sea.

With evening approaching, we decided to go our separate ways. I waved goodbye to the Hurricane Intercept Research Team and set off to find a cheap motel, leaving Mark Sudduth already itching to check out another tropical disturbance mobilising in the Gulf of Mexico. We were all finished with Gustav, although the storm itself still had plenty of life left: in less than twenty-four hours Gustav would be a full-blown hurricane, the first of the 2002 season, eventually reaching a sustained windspeed of 145 kilometres per hour as it tracked rapidly northwards towards Canada. All I saw of it next morning were delicate fingers of cirrus cloud, backlit by the sunrise – the outer fringe of the new hurricane as it spun far out in the Atlantic.

FLORIDA

Having experienced a tropical storm, I now needed to find out how these powerful cyclones might already have been affected by global warming. So I had booked an interview with James Elsner, a hurricane expert at the Department of Geography in Florida State University, based in the state capital Tallahassee.

I got from Washington DC to Florida on a Greyhound bus, through a succession of small Southern towns: Fayetteville, North Carolina, then Santee, Walterboro, and Ridgeland, South Carolina. Each place looked identical – with strip malls lining the road into town, joining the ubiquitous gas stations, auto warehouses, fast-food joints and the occasional church in a procession of numbing sameness.

But even as the urban geography remained the same, the

air around was changing – becoming warmer and more humid, with towering clouds packing the occasional tropical downpour. We reached Brunswick, Georgia, and away from the strip malls there were more Southern-looking houses along the road – rickety wooden structures with fading paint, easy chairs swinging under the eaves of their front porches.

Tallahassee is on the Florida Panhandle, where the state juts out along the Gulf Coast underneath Georgia and Alabama. It's the part of the 'sunshine state' that most tourists never see, where the official comings and goings of the state legislature take place an uneasily short distance from the harsh (and often racist) conservatism of the Deep South. I booked into a soulless but cheap Super 8 motel along a dual carriageway heading out of town. Here, as in all of America bar a few crowded city centres, the car was king: there was no footpath between my motel and the Chinese restaurant next door, and I had to leap across a grassy drainage ditch to get from one driveway to the other.

It rained on and off most of the night: by peculiar coincidence another tropical storm, this one called Hanna, was sweeping in off the Mexican Gulf. Hanna was much weaker than Gustav, but ten centimetres of rain had already fallen over much of the Panhandle and warnings of floods were being broadcast on the Weather Channel.

James Elsner showed up on the dot of ten o'clock the next morning, in the midst of a monstrously heavy rainstorm. We were soaked through in the twenty seconds it took to get from the motel entrance hall over to the car.

'We're gonna get some flooding,' Elsner commented evenly as we pulled away, the rain pounding on the roof and sparking enormous muddy-brown torrents on both sides of the road. We dashed from the car to his university offices under a shared umbrella, splashing through widening puddles as the rain poured down from a dark sky.

Elsner's main work is focused on the historical climatology of Atlantic hurricanes – looking at the storms of the past to try and pin down any identifiable differences that might betray the fingerprints of global warming. Catastrophic hurricanes are not new: pictorial engravings left by the Mayans in Mexico provide the earliest human record of their destructive power over a thousand years ago. Some of the first European casualties of an Atlantic tropical storm were a shipload of passengers heading to the new colony of Virginia who were struck by a powerful hurricane in 1609 – forcing them to land in uninhabited Bermuda where some of them became the island's first permanent residents.[1]

Maps of old storms were plastered across Elsner's office walls, including one that hit Apalachicola in 1850 and the 'Great Gale' of 1804, which destroyed Fort Green on the Savannah River, drowning thirteen people. Although there's no guarantee that every storm, especially those far out at sea, would have been recorded in the days before aircraft reconnaissance and Earth-orbiting satellites, there are extensive historical records available from ships' logs and old newspaper reports. 'In fact, 1837 sticks out as a notorious year along this part of the coast of Florida, from

Tallahassee westwards to Pensacola,' Elsner told me, 'we got hit six times by tropical storms, three of which were hurricanes, and all in one year.'

So had the number of hurricanes hitting the US changed – perhaps increasing in the twentieth century because of global warming?

'Not as far as we know. With overall activity in the Atlantic basin we've seen active periods and inactive periods, and there's very little evidence to suggest that the twentieth century was any different from the nineteenth century. So even if you look at the group of years starting in 1901 and ending in 2000 and compare it with the group of years starting in 1801 and ending in 1900, statistically they're really quite indistinguishable overall.'

Until quite recently the trend in US hurricane landfalls has actually been downward, something scientists attribute to natural oscillations in Atlantic currents. Between 1947 and 1965, for example, fourteen major hurricanes struck the East Coast, but between 1966 and 1983 there were none.[2] The years 1991 to 1994 were also remarkably inactive, with the lowest frequency of intense Atlantic hurricanes for any four-year period since records began.[3]

But in 1995 everything changed. That year saw nineteen named storms – over double the average – eleven of which were hurricanes and five evolving into 'intense' hurricanes. It was the second largest number of named storms of any year since 1871 (the largest had been in 1933 with twenty-one storms).[4]

And that was just the beginning. Every year between 1995

and 2001 (with the single exception of 1997) has seen above-average numbers of Atlantic hurricanes. In fact 1995 to 2000 saw the highest average number of major hurricanes for any six-year period since 1944, when reconnaissance aircraft flights began to make records more reliable.[5]

Overall, during the last eight hurricane seasons the Atlantic seems to have shifted into hyperactivity, spawning many more storms, some of them of terrifying intensity. Given that the 1990s had been the hottest decade in recorded history, could this be a sign of global warming?

'Well,' answered Elsner hesitantly, 'if you want to look at *global* warming you've got to look at *global* tropical cyclone activity, of which the Atlantic represents only about ten per cent. The problem here is that reliable data only goes back to the 1960s. But when you look at it, you still don't see any trends.' In fact, there's actually something of an anti-correlation between the Atlantic and other areas: although 1995 saw a near-record high point in tropical cyclone activity in the Atlantic basin, the eastern North Pacific by contrast saw a near-record low, and the global number of tropical storms and hurricanes was also below normal.[6]

Other tropical ocean areas also betray very few signs which might point to definite trends: in the western North Pacific there's been a decade-long rise in tropical cyclone activity, but this was preceded by a fall of similar magnitude.[7] The Australian region has also seen a downward trend since the 1970s.[8] But the Pacific islands seem to be experiencing more storms – when I visited Tuvalu everyone was complaining that tropical cyclones were more frequent than ever before.

'I can show you the numbers,' Elsner continued. 'We can sit here and count up the number of storms around the globe and group them by years. And you can look at that number and you don't see it moving anywhere. Yeah, it goes up and down, but it's *not trending*. If it was trending, it would be reported all over the scientific literature. Scientists are always the first ones to tell you if something's happening.'

But Elsner hadn't quite finished. 'Now there is one aspect that may be of interest to you with regard to global warming and hurricanes,' he went on, leaning forward with a slightly conspiratorial air. 'And that is that in the 1990s there wasn't *a single year* with the normal average number of hurricanes. If you look at a graph you see a tendency towards more extreme years: meaning that it's either very active or very inactive. So the climate either conspires to produce lots of storms or shuts it off almost completely. That's an interesting concept of global change that may be occurring.'

'So it's like the system is becoming more chaotic?'

'The system doesn't seem to be as stable, there's more variance. And the insurance industry is very interested in this – they use the word "volatility", like with the stock market. This is curious to me from a scientific point of view.'

It had stopped raining by the time we left the building. Elsner gave me a lift back to the bus terminal – on the way we talked about the latest tropical depression to form in the

Caribbean. 'I'd venture a storm developing and maybe even hitting Florida later in the week. That might be worth your while hanging around for.'

I did hang around for three days in Miami whilst Tropical Depression 10 made halting progress across the north coast of Venezuela and then went back out to sea towards Cuba. On the day I visited the Hurricane Research Division of the US government's National Oceanic and Atmospheric Administration (NOAA) it was the main topic of conversation.

The HRD is a large white building offshore from downtown Miami on Virginia Key, over the road from the Miami Seaquarium. It is built of a tough poured concrete expressly designed to stand up to the most powerful hurricane imaginable. On the second floor the HRD's top scientists were dragging chairs in from other rooms and clustering around two large screens, one of which was looping an animated satellite picture of the revolving tropical depression.

'The system we're most interested in is what was TD10,' began Ray Zehr, a visiting climatologist from Colorado State University. 'Yesterday the centre lost much of its convection. However, overnight, you can see that it has changed and now looks very different from before. The question is whether it's a new one.'

'That's a new one, it's a totally new centre,' one of the scientists interrupted excitedly, sparking off a discussion about whether the system would need a new name or would still be classified as TD10.

'All right, but what's it forecast to do?' came an impatient question from somewhere on my right.

'Okay . . . the forecast,' continued Zehr. 'All the models take it south of Jamaica and then over western Cuba. Then most of the models drift it upwards towards south Florida. The AVN takes it further east. But, anyway, over the next three days there's a good possibility of a strong tropical storm or even a minimal hurricane drifting north between Jamaica and Grand Cayman.'

'What's the shear situation?' another questioner asked. (Shear is upper-level wind which can cut the top off a developing hurricane and stop it gaining strength.) TD10 would probably only be affected by minimal shear, Zehr replied, potentially allowing it to intensify into a stronger system.

After fifteen minutes more of detailed discussion about the atmospheric conditions surrounding TD10 in the Caribbean, as well as tropical cyclone activity in the eastern Pacific, the meeting was drawing to a close. The scientists were briefed on which 'hurricane hunter' aircraft would be flying across the Caribbean to investigate the system – one was already on its way.

'So, what's your personal forecast, Ray?'

'What's my forecast? Oh, I disagree with the models. I'll take it to Cancun as a hurricane,' Zehr replied with a grin. I couldn't tell whether this was a joke or not. But oddly enough, he would turn out to be right.

The Hurricane Research Division's director Hugh Willoughby had been sitting quietly on my left. Willoughby manages to combine both geniality and shrewdness – important skills for someone who had not only pioneered the scientific study of hurricanes, but also – as HRD's director – had managed the bureaucracy, budgets and staff for the previous seven years. He had recently 'retired' back into real science, and was consequently in a particularly jovial mood.

'We're a bit wary of writers at the moment,' he grinned. 'The last guy who came here – I think he wrote a book on Hurricane Mitch – described one of our best scientists as "resembling a manatee".'

'Was that the guy two seats to my right? . . . Oh, he didn't look *anything like* a manatee.'[9]

Until recently Hugh Willoughby had been a world record holder, having flown into hurricanes 416 times. He had been on the HRD plane that flew into the 1989 storm Hurricane Hugo at an altitude of only 450 metres, a decision made without realising that Hugo had just intensified from a Category 2 to a Category 5 – the most dangerous storm possible on the 'Saffir/Simpson' scale.

He described later how everything in the aircraft was thrown around and turned upside down: a twenty-man life raft strapped in front of his seat broke loose and hit the ceiling. Coke cans, coffee, paperwork and peanut-butter sandwiches were sent spinning wildly all over the aircraft. At one point he saw his laptop flying through the air in front of him. When they eventually broke through into the eye of the storm, one of the engines had caught fire. Desperately turning to stay within the calm eye, the stricken plane was at one point just thirty seconds away from hitting the sea.

'It's a little strange,' Willoughby reflected afterwards, 'to be going round and round in the eye thinking: this may be the afternoon that I die.' But he hadn't prayed: 'We got ourselves into that situation. Asking God to get us out of it seemed a little presumptuous.' Instead he had forced himself to fly back into the same hurricane – just as someone who falls off a horse needs to get straight back in the saddle and carry on riding, or they'll never do it again. But for years afterwards, every time his hurricane hunter aircraft hit turbulence, Hugh Willoughby admits that he broke out in a sweat.[10]

Perhaps his military training helped him keep a sense of detachment: trained as a naval meteorologist, Willoughby was in California when Hurricane Camille flattened the Gulf Coast. 'They were looking for people with good academic backgrounds to ride in the recon planes. And I said: "That sounds like fun!" '

Hurricanes were a primarily academic concern to the scientific staff at the Hurricane Research Division until 1992, when Hurricane Andrew hit Florida packing winds of up to 264 kph: a borderline Category 5.[11] Many staff members had their homes damaged or even destroyed: one of Willoughby's colleagues, Stan Goldenberg, had to shelter under a mattress with his family as his house collapsed around him.

Willoughby remembers gathering his whole family by torchlight into one room on the lee side of the house where they rode out the storm, listening to the wind shrieking above them like an express train. At HRD headquarters the whole ground floor was flooded. The exact top wind-speeds are still unconfirmed, as most measuring equipment was simply blown away. The National Hurricane Center was able to take one last radar image, showing Andrew's eyewall just making landfall to the southeast, before the radar was blown off the roof.

'The storm was extremely small, with a very sharp gradient of the wind outside the eye, so literally the damage pattern changed from block to block as you went south. What would have happened in more extreme conditions to my house is that you would have started hearing windows fail and, as the wind came into the house, there would have been failure of the internal walls. And the really scary thing that happens if you breach into the attic is that it'll blow the ceiling down from above, which is not good if you're under it. I mean, it comes down with a lot of force. That would kill a person.'

Twenty-six people in total were killed by the storm, which crossed Florida into the Gulf of Mexico and then made landfall a second time in Louisiana, and 100,000 homes were damaged, 25,000 destroyed completely. Damage estimates put the bill at around $25 billion, and the death toll would certainly have been higher had not a million people been evacuated in Florida.[12] In fact, the worst of Andrew narrowly missed Miami's downtown area – if the eye had passed a few miles further north the damage cost would have been vastly greater.

Minimising the death and destruction caused by hurricanes is HRD's main preoccupation. And the strategy of improving hurricane preparedness and mounting full-scale evacuations when a storm approaches has been remarkably successful. As Willoughby put it: 'You can show clearly that your odds of dying in a hurricane if you live on the Gulf Coast or East Coast of the United States have decreased by two orders of magnitude since 1900: it's basically an exponential decrease. And you could argue that we're actually at an irreducible minimum: the people who die are unfortunates – they exercise bad judgement or are extremely unlucky.'

But look at damage statistics and the trends point the other way. 'Property damage has gone up like crazy,' Willoughby acknowledged. It's tempting to ascribe the increase in damage to climate change – and many have indeed made the connection[13] – but Willoughby wasn't convinced. 'The hurricane damage statistics are driven entirely by economic factors – there's just more stuff sitting around on the beach waiting to be blown away or washed away.'

His point is backed up by a recent study co-written by one of his HRD colleagues Chris Landsea, which found that the rising hurricane damage totals were not due to stronger storms, but rather to a massive increase in people and property located along vulnerable US coastlines. (For example, more people live in Dade and Broward counties of south Florida today than lived in the whole southeastern United States in 1930.)

To make sense of the trend, the authors 'normalised' the

damage data: adjusting for inflation, population increase and the fact that wealthier people simply have more possessions to lose when a hurricane moves ashore. Landsea then found that the graph of damages pointed the other way, and the 1940s, 1950s and 1960s had more frequent and costly hurricane landfalls than the 1970s and 1980s. Hurricane Andrew was still the second most costly storm – but the other top five all occurred before 1950.[14]

But nor can a climate change connection be ruled out. Although it's true that damage statistics are notoriously misleading, climate science is beginning to show some real changes in the nature of Atlantic hurricanes. A paper published in *Science* in July 2001 by several top hurricane researchers – including two HRD experts – warned of an 'apparent shift in climate' in the Atlantic, which together with recent increases in coastal populations could result in a 'catastrophic loss of life' in the event of an incomplete evacuation ahead of a rapidly-intensifying storm. The authors also allude to the possibility of a global warming fingerprint, writing that the unprecedented increase in Atlantic hurricane activity since 1995 has in part been driven by 'the additional increase in sea surface temperatures resulting from the long-term warming trend'.[15]

Willoughby also thought that he might be seeing something new. 'There's a cyclic variation in the number of the most intense hurricanes that's basically twenty to thirty years of more hurricanes and twenty to thirty years of fewer hurricanes. So there was a suppressed cycle that started in about 1970 and lasted through 1995, and then an active cycle which began in 1995 and, apart from El Niño, is still going on.' (El Niño, a warm-water current that occasionally crosses the Pacific from west to east, tends to suppress Atlantic hurricane activity by increasing wind shear, and thus cutting the developing storm clouds in half.)

'But this current active phase appears to be a little bit more active than the previous active phase. It's not dramatically so – just a few per cent – and it may be that we're just observing more comprehensively but, if it is more active than the previous active phase, then that might be a global warming signal.'

And, although there's a strong current of climate change scepticism running through the tropical meteorology community, Willoughby was no longer a subscriber: 'I've gone from being a lot more sceptical to thinking that it's something we really need to be concerned with.'

Part of this was straightforward theory: simply, that a warmer ocean would have the potential to spark stronger storms. A hurricane is basically a heat engine, and by measuring the sea temperature underneath a developing storm you can use the laws of thermodynamics to calculate the maximum potential intensity the storm can attain. And with a warming ocean there wouldn't necessarily be more storms, but 'what's going to happen is that the maximum potential intensity is going to creep up'.

Even as we spoke, Tropical Depression 10 was intensifying – a day later it had reached tropical storm strength and been given the name Isidore. Intensely heavy rains were already falling over much of Jamaica, and a 'tropical storm watch' was posted for the Cayman Islands. Two days later the National Hurricane Center put out its 11 a.m. bulletin, headlined: 'ISIDORE NEARING HURRICANE STRENGTH . . .', by 5 p.m., Isidore officially became the second hurricane of the 2002 season.

Tropical storm watches were posted for the Florida Keys: the forecast was still for the hurricane to move over Cuba and then threaten the Sunshine State. Western Cuba did indeed get a battering – with sixty centimetres of rain and a storm surge of up to three and a half metres as well as hurricane-force winds, forcing the evacuation of 250,000 people – but then Isidore veered west towards Mexico's Yucatan peninsula and continued to strengthen.

On 21 September the NHC was sending out the stark warning: 'ISIDORE STRENGTHENS INTO A MAJOR HURRICANE OVER THE YUCATAN CHANNEL AND EXPECTED TO GET STRONGER . . .' By 4 p.m. the next day Isidore's powerful eyewall was battering the Mexican coast, packing winds of 200 kilometres per hour. About 70,000 people were evacuated from low-lying fishing villages along the coast, and the colonial state capital Merida found itself ankle-deep in water whilst trees were uprooted, power cut and roads and homes flooded across the state. Four people were killed in road accidents related to the storm.

And although Isidore lost power as it moved across dry land, late on the 24th its centre was back over the warm ocean. By this time it was so large that the circulation covered most of the Gulf of Mexico – the first bands of squally rain began moving onto the southern US coast even as Merida was still being pounded. Hurricane watches were posted for Louisiana and Mississippi as the storm approached – but Isidore never regained its former power, eventually making landfall just west of New Orleans whilst still only classified as a tropical storm. As the city breathed a sigh of relief that Isidore's winds had remained below hurricane strength, torrents of rain were falling, leaving flood pumps struggling to keep up. But New Orleans – which lies below sea level and is probably the US city most vulnerable to a major hurricane strike – had been spared once again.

Even as New Orleans celebrated, another tropical storm called Lili was forming around Jamaica, and this one *would* threaten the US coast. Initially following almost the same path as Isidore, Lili also pounded western Cuba with heavy rains and strong winds, forcing 330,000 people to move to safer ground. Hurricane warnings were again issued for the US Gulf Coast – this time all the way from Texas to the mouth of the Mississippi River.

By Wednesday 1 October the National Hurricane Center was sounding worried. At 1 p.m. it warned that Lili had suddenly strengthened into a 'DANGEROUS CATEGORY 4 HURRICANE', advising that 'PREPARATIONS TO PROTECT LIFE AND PROPERTY IN THE HURRICANE WARNING AREA SHOULD BE RUSHED TO COMPLETION . . .' Three hours later a new advisory upgraded Lili to an 'EXTREMELY DANGEROUS CATEGORY 4' storm, which was expected to pack 'A LIFE-THREATENING STORM SURGE OF TEN TO TWENTY FEET [three to six metres] ABOVE NORMAL TIDE LEVELS'.

Residents of coastal Texas and Louisiana needed no more warning, and nearly a million of them headed inland as the strong hurricane barrelled towards the shore. Throughout the region people boarded up their houses and offshore oil platforms were evacuated. The Red Cross set up emergency command centres as the NHC issued final dramatic warnings. Just ten hours before Lili was due to make landfall, winds were up to 230 kilometres per hour – the kind of speed that would cause catastrophic damage to buildings and enormous flooding.

But it didn't happen. Inexplicably, the hurricane suddenly lost power as it approached land, with its strongest winds only a little over 140 kilometres per hour. Three people were hurt by debris and falling trees, but no one lost their life. 'It looks like we were lucky,' declared Louisiana Governor Mike Foster. The National Hurricane Center's director Max Mayfield was at a loss to explain Lili's downward transformation. 'A lot of PhDs will be written about this,' he said.[16]

NEW JERSEY

Whilst Hugh Willoughby and his colleagues at the Hurricane Research Division and the National Hurricane Center fight a daily battle to better understand and forecast real-life storms, other climatologists in faraway New Jersey work with storms that exist only within a computer. One of these experts is Tom Knutson, a leading climate modeller at the Geophysical Fluid Dynamics Laboratory in Princeton, whose work looks exclusively at the future, examining the potential for stronger tropical cyclones in a globally-warmed world.

One of Knutson's main successes, in collaboration with Bob Tuleya, has been in improving the computer modelling of hurricanes used in global climate change research. This has long been a headache for researchers: because models of the global atmosphere work on grid squares of several hundred kilometres, hurricanes are simply too small for their intense revolving cores to show up properly.

Knutson got around the problem by 'nesting' a higher-resolution regional hurricane model within a global climate simulation, so that tighter-packed grid squares could more accurately resolve the virtual hurricane's eye, whilst the global model outside could provide the wider atmospheric conditions surrounding the storm. Then, after checking his virtual representation to make sure that it represented real-world hurricane intensities reasonably well, Knutson took a globally-warmed future a century from now as the baseline and looked at the behaviour of the computer-generated storms within it.[17]

This took a staggering amount of computing power – in their latest experiments, numbers were crunched on a massive supercomputing system, using hundreds of processors simultaneously – in order to resolve the three-dimensional model, which took nine kilometre grid points of the atmosphere through forty-two vertical levels over a 500 kilometre by 500 kilometre region. In some experiments the atmospheric model was even coupled with an interactive ocean to give a more realistic simulation of the real world.[18]

The results were compelling: 'What we found was that in the high-CO_2 climate the storms were stronger – something like five to ten per cent stronger in terms of intensity.' The model was clearly indicating that higher tropical sea surface temperatures increased the potential power of each storm.

And if 5–10% doesn't sound like much, remember that hurricane damage goes up exponentially with the wind-speed: so with a doubled windspeed damage increases by four times or more, with tripled windspeed it would increase by nine times or more. Indeed, as Knutson pointed out, 'Because the damage goes up at a faster rate than the intensity, there could be more like a ten to twenty per cent, or perhaps even greater, increase in damage potential.' This conclusion did not apply just to the Atlantic, either. 'We looked at conditions for all the tropical storm basins, and they were all indicating this potential for stronger tropical cyclones.'

'But given that sea temperatures have been rising for decades, why hasn't this effect *already* been observed?' I asked.

'Well, over the last fifty years global sea surface temperatures have warmed a few tenths of a degree, but I don't think that sea surface temperatures in the tropical storm basins with the best observational records of hurricanes have warmed even as much as the global average, or in the case of the Atlantic basin have even warmed at all. So the relevant sea surface warming has been less than half a degree typically, during the period of relatively reliable hurricane observations.' Given that Knutson's model got a 5–10% increase in storm intensity with a 2.2°C rise in tropical sea temperatures, the less than half-degree rise already observed 'translates into, what, maybe a one per cent increase in intensity. You just can't measure that accurately.'

This explained rather neatly why no one could find a clear global warming signal in hurricane intensity trends so far – because you simply wouldn't expect it.

But the future was a different issue, Knutson emphasised. Even without global warming, 'just if we repeat the experience of the twentieth century – the amount of development that has occurred in these hurricane-prone areas means that there are a lot of disasters waiting to happen, regardless of the climate change effect we're talking about'.

However, the additional global warming effect would 'add a little more salt into the wound, I think. You know, if Hurricane Andrew had had ten per cent higher windspeeds that could really have caused some major additional damage, I'm sure.' And that's without factoring in sea level rise, which would increase the destructive power of storm surges and flooding. 'If these predictions of future warming, with increased hurricane intensities and sea level rise are true, we haven't seen anything yet – especially with the increase in population and development in hurricane-prone regions.'

I wondered what the victims of Hurricane Mitch – who probably thought that their experience was as bad as it gets – would think of that. A metre of rain fell in some parts of Honduras between 25 and 31 October 1998,[19] sparking catastrophic floods and mudslides. An estimated 60% of the country's transport infrastructure was damaged – with 107 major roads cut, 189 bridges destroyed and eighty-one towns cut off.[20] In the Honduran capital Tegucigalpa, the third floor of a hospital had to be evacuated as rising flood-waters began to pour into the second floor.[21] In neighbour-ing Nicaragua 2000 people were killed in a single mudslide from the Casitas volcano on 30 October. A year later new downpours began to turn up skulls from the thick layer of mud.[22]

One woman was washed eighty kilometres out to sea, but was miraculously rescued by helicopter after spending six days clinging to wreckage. Her story was one bright spot amidst appalling tragedy. According to the 1999 Red Cross *World Disasters Report*, 10,000 people were killed in total, and the affected Central American countries suffered $7 billion in economic losses, only 2% of which was covered by insurance.[23] As the Honduran President put it: 'We lost in seventy-two hours what we had taken more than fifty years to build.'[24]

But according to Knutson's work, the damage from a future storm like Mitch could become even worse. Although storm intensity and windspeed might only go up by 5–10%, his models revealed that rainfall could actually surge by up to a third. 'We found that the precipita-tion near the storm was quite sensitive to the warming – it increased by about twenty per cent. In some experiments the maximum precipitation rate anywhere within the storm increased by over thirty per cent. We think what's going on there is that in the warmer climate the atmosphere's holding a lot more moisture. And the more moisture the air is holding, the more water is available to be rained out by the storm.'

I was shocked. 'Imagine you were under the heaviest part of Hurricane Mitch, and you got thirty per cent more rain . . .'

'That's right,' he agreed.

Although Hurricane Mitch was so extreme that 'Mitch' joins Georges, Hugo and Andrew on the list of infamous names that can never again be used to identify a hurricane, it was still less deadly than an even more horrific disaster which struck Bangladesh in 1991. Storms don't get names in the Indian Ocean, so the cyclone that killed 138,000 on 29 and 30 April of that year will only ever be known as Tropical Cyclone 02B. It was a Bay of Bengal 'super-typhoon', which packed windspeeds of up to 260 kilometres an hour at its most intense[25] – equalling both Mitch and Andrew as a Category 5 giant, and becoming one of the deadliest storms in recorded history.

Bangladesh has always been particularly vulnerable to cyclone disasters. Most of the country's dense population is packed into the delta of the Ganges and Brahmaputra rivers, many of them scratching a living on shifting offshore islands, all on land barely two metres above sea level. Tropical Cyclone 02B threw up a storm surge of eight metres in places, the gigantic tidal wave aggravated by the triangular funnel shape of the Bangladeshi coastline and the fact that the storm struck at high tide.

One woman, who lived near the coastal town of Chittagong, spoke of a 'wall of water' rushing towards her home. 'The ground shook and the skies split with a roar so loud that I thought I had gone mad,' she remembered later. She had just managed to wrap a rope around her three children when the wave broke over their heads, and they spent the next eight hours clinging to a post on her roof, before the house was washed away and the family plunged into the floating debris. Somehow she and her sons survived, but her husband – who was away at the time – never returned.[26]

Another survivor in Chokoria district – normally a tranquil place of water channels, coconut palms and shrimp pools – described spending the night tied with his family to the strongest tree he could find, whilst the wind and waves pounded them with such fury that most of the clothes were ripped from their bodies.[27] Another man was not so lucky: his wife and five children were carried off by a six-metre-high tidal surge.

Many of those unable to reach concrete cyclone shelters were swept away – often as they crowded outside trying to get in – as entire offshore islands were submerged. On Sandwip Island alone 23,000 people were drowned; many of the survivors were packed into a mere eight cyclone shelters. Most people had heard the storm warnings hours before, but following a rash of false alarms in previous years, simply didn't believe them.[28]

Two days later rescuers in helicopters were unable to land because the islands were still under knee-deep water. Amongst the ruins of the mud and thatch houses, human corpses and animal carcasses floated in grotesque rafts. Survivors were desperately turning the corpses over, looking for disappeared family members. The stench rapidly became overpowering, and cholera began to break out.

One witness described surveying the area from the air, finding that 'entire islands and their populations have now disappeared, and others, unpopulated except for corpses, have emerged in the silt'. The pilot attempted in vain to pick out villages from his map. 'Through binoculars one could see a thousand bodies, some swaying lazily with the current, their soft parts distended, their bellies swollen by bacteria.'[29]

The port at Chittagong was devastated: twenty ships sank in the entrance channel, whilst cranes toppled over and crashed down onto storage sheds, destroying everything inside. One floating crane broke anchor and ploughed into a newly-built bridge, destroying the middle section. An embankment around the airport gave way, and several military aircraft and helicopters were left piled on top of each other in a lake. A quarter of the region's entire agricultural product, industry and physical infrastructure was wiped out, the damage cost totalling $1.3 billion.[30]

In the control room at the Joint Typhoon Warning Center on faraway Guam, forecaster Stacy Stewart was watching the whole thing happening, issuing desperate last-minute warnings as the super-typhoon hurtled towards the fragile Bangladeshi coast. Once it hit, there was nothing else he could do. 'When the storm came onshore in satellite imagery, we all knew that a terrible tragedy was unfolding before our eyes and we all wept,' he recalled later.[31]

Hurricanes Andrew, Mitch and Tropical Cyclone 02B – together with an even stronger storm which hit the Indian coast at Orissa in 1999, killing 10,000[32] – belong to a freak category of super-intense tropical storms that are still mercifully rare today. But it is these super-storms which seem due to appear ever more frequently as sea temperatures creep up over the coming decades, delivering increasing power to these deadly atmospheric heat engines. And the impact of these cyclones – from Bangladesh to Florida to Tuvalu – will get increasingly worse as sea level rise increases the height and, consequently, the destructive force of the storm surges which accompany them.

If Tom Knutson's computer models are right, and 'we haven't seen anything yet', then many more people will lose their lives to hurricanes, tropical cyclones and typhoons as global warming accelerates during the century ahead.

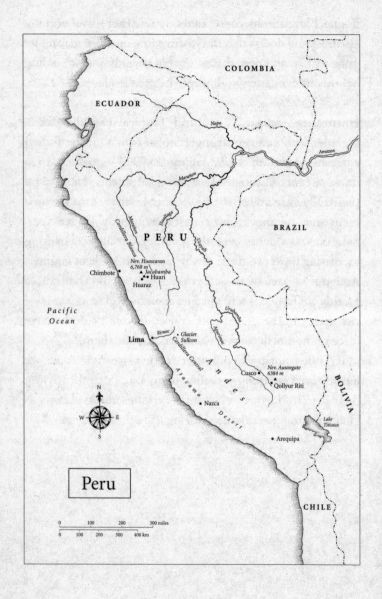

Peru

6

Peru's Melting Point

Lima was both strange and familiar. The same hand-painted former school buses plied their way up and down the wide Avenida Arequipa, their boy conductors leaning out into the traffic to yell the destination at passers-by. The side streets still smelt of urine and rotting vegetables, whilst the whole city centre was covered more thickly than ever with the familiar layer of grime.

A good rainstorm would have washed the whole lot away – but Lima, stuck in one of the driest deserts in the world, never gets rain. Above the whole city rose Cerro San Cristobal, the 600-metre hill we had once climbed with the British ambassador – who, convinced that someone would try to steal his expensive wristwatch, had a handgun stowed carefully away in his trouser pocket.

The changes were less perceptible: a newer look to the cars, and taller, more modern buildings, the kind of things you'd expect to see after a twenty-year gap. The pervasive smell of fish seemed to have lifted too, perhaps because of coastal overfishing, which has left Peru's principal export industry in decline. Gone too was the insidious fear of terrorism: when my family left in 1982 the country was in near civil war, with bombings, kidnaps and murders – the former perpetrated by the Sendero Luminoso ('Shining Path') guerillas, the latter by guerillas and the army in equal measure – tearing the social fabric apart. The building next to my father's office was blown up one lunchtime, killing a woman caretaker, and whilst travelling in the interior we once missed being caught in a shoot-on-sight curfew in the highland town of Ayacucho by a mere five minutes. I can still remember hearing the rattle of gunfire as we cowered in our hotel room behind heavy wooden gates.

I couldn't resist a quick visit to my old school. Markham College was an embarrassingly exclusive place intended only for foreigners and rich Peruvians – the high fees for which were paid during the three years I attended (thanks to my father's employment in a geological overseas aid project) by the British government. The deputy headmaster, a genial man of about fifty who kindly pretended to remember me, called over two impeccably-dressed prefects to show me round. (The word 'prefect' brought a shudder of recollection, of being caught out of bounds by an enormous prefect with a black armband, who'd reduced me almost to tears with the threat of a fearsome 'black star'.) The three of us paced self-consciously around the grounds together. It all looked impossibly small, especially the playing field – which I remembered as an enormous grassy prairie so huge I'd never even reached the other side of it. No one was playing marbles any more either, which seemed a shame.

Next I dropped in on my family's old house, and stood outside for ten minutes whilst the maid listed various reasons why she couldn't possibly let me in. Before she slammed the door, I peered over her shoulder and got a brief glimpse of the front garden. Everything was wrong – instead of the homely chaos I remembered, it all looked impeccably neat and tidy. Then (for the first time) I noticed how much the whole area was fortified, with all the customary high walls, security cameras and even electric wires necessary to keep the countless millions of poor from bothering the few thousand rich, and was glad that we'd moved on.

I was staying in a hotel in the grimiest part of Lima's grimy city centre: a rambling former colonial mansion with tangled vines climbing up every wall and a collection of mouldering skulls in a downstairs cabinet. My arrival left the entrance hall full of climbing boots, sleeping bags, food and cooking utensils, all propped up haphazardly against the hotel's bizarre collection of half-built mock-classical statues. Most of the gear belonged to my travelling companion Tim Helweg-Larsen, an enthusiastic graduate I'd recruited for his mountain-climbing skills. In the inaccessible places I aimed to go, solo climbing is simply too risky: even the smallest accident could turn into a death sentence.

Tucked away most safely of all at the back of my rucksack was a folder of A4 colour prints, selected by my father and me back in Wales, of the Andean glaciers he'd worked on in 1980. I stole a look at them every now and then, especially the one with the fan-shaped glacier above the lake – even though I already knew every inch of that imposing wall of ice. These photographs were vital. Without them, I would not be able to answer the question which was the central objective of my journey: what had happened to that lakeside glacier?

In the process I also aimed to find out much more about how glacial retreat was viewed by ordinary people, and about how the rapid disappearance of some much closer glaciers was endangering the survival of Peru's capital city itself.

QOLLYUR RITI

The photographs were taken in the Cordillera Blanca, a range of ice-clad peaks four hundred kilometres north of Lima in the highest part of the Peruvian Andes. But before heading up there, I wanted to make a detour into the south of the country. I had heard a historian on a discussion programme on Radio 4 a few months before talking about how one of the traditional Andean religious festivals was held annually at Qollyur Riti – an inaccessible and freezing sacred valley.

Local people would converge there and make pilgrimages up to the glaciers to cleanse themselves of sinfulness and – a little covertly, since Peru is a Catholic country – to worship the old mountain gods of the Incas. The historian had gone on to mention, in passing, that on a recent visit people had been talking about how the sacred glaciers were retreating back up the mountains. Did this mean that the mountain gods, known as Apus, were leaving them?

He left the question unanswered, and the conversation moved on. Meanwhile, I grabbed a piece of paper, scribbled down the name, and began some research.

Cusco was just as stunning as I'd remembered it. As our bus wound down into the valley, its two main plazas seemed to swim in a sea of red-tiled roofs. Every corner and alleyway held some fabulous new discovery, like a rare Inca snake carving or a hummingbird feeding on a tree covered with immense orange and white flowers. Cusco had been the centre of the Incan empire which, before the Spanish conquest, stretched right from modern-day Ecuador to central Chile, and encompassed pretty much all of the Andean mountains, most of the desert coastal strip, and much of the eastern Amazon jungle besides.

On the hillsides above Cusco lies the enormous fortress of Sacsayhuaman, a stepped rampart of zigzag walls built out of immense stone boulders. The fortress was the site of one of the most critical battles of the Spanish conquest, when thousands of Inca warriors almost succeeded in recapturing their capital after a prolonged and bitter siege. Only a last-ditch charge by fifty armour-clad Spanish cavalry defeated the Inca resistance.

The Spanish pillage was so meticulous and thorough that barely a scrap of Inca gold survives today – the whole lot, whatever the artistic significance of the work produced by the skilled Incan goldsmiths, was melted down into bullion and transported in galleons back to Spain. But enough of Cusco survived the *conquistadors'* onslaught to give a good idea of how splendid it must once have been. The lower parts of many walls in the town are still composed of finely-worked Inca stone blocks, which have survived untouched down the centuries even whilst frequent earthquakes repeatedly destroyed the shoddy colonial masonry standing above them. (And the colonial buildings too have their own gentle, faded charm.)

Just off the main plaza is one of the finest walls of all, once part of the palace of the sixth Inca emperor, Inca Roca, and now one side of a museum. Halfway along is a large block carved with a scarcely believable twelve sides. Like the others, it fits snugly into its neighbours without so much as a millimetre on either side – masonry work unmatched anywhere in the world.

And the stonemasons who constructed it took their secrets with them: to this day no one knows how they managed to achieve such perfection. Even our hotel had an Inca doorway which gave me a thrill of architectural pleasure every time I passed through it.

But we couldn't afford to hang around. Up in the high mountains the Qollyur Riti festival was already well underway, and after asking around I discovered that the dramatic pilgrimage finale would be happening in just two days' time. Tim was wary about an immediate departure to high altitude, leafing through a book he'd brought with him about the dangers of Acute Mountain Sickness.

'It says here that you're supposed to take a day to adjust to every three hundred metres of altitude you gain.'

I did a brief calculation. 'But that's ridiculous. We're at three thousand metres here, and Qollyur Riti's at nearly five, so it's supposed to take us, what, six days?'

'Well, that's what it says. We should at least spend a day at the village though before making the trek up to the valley.'

'Maybe,' I agreed. 'But the worst that happens in my experience is that you feel a bit sick. You get a headache or even throw up. Then you get better.'

We did indeed both feel sick the next morning as the bus wound its way up several dozen hairpin bends towards a pass. Dust clouds billowed behind, coating everything inside the vehicle. Our fellow passengers – all colourfully-dressed *campesinos* in ponchos and hats, grinned and nodded, offering us snacks of baked broad beans.

There seemed no shortage of food: at almost every stop, women would crowd around, trying to sell dirty boiled potatoes and roast guinea pig through the bus windows. Neither of us felt like eating though, and when the bus stopped at the top of the pass next to some roadside shacks on a bleak-looking plateau, my legs almost gave way as I staggered down the steps. Alpacas and llamas were grazing on the coarse mountain grass, watched over by children in tattered clothes. In the distance, a range of ice-clad mountains gleamed impossibly white on the horizon, topped by the sacred peak Nevado Ausangate. Qollyur Riti, I knew, lay somewhere nearby.

The bus broke down twice on the far side of the pass, but since all Peruvian bus drivers – and a good many passengers – are skilled mechanics, it only set us back a couple of hours. It took even longer to get through a narrow gorge on the one-lane dirt road, as returning trucks and minibuses kept appearing around blind corners, resulting in much reversing and swearing amongst even bigger clouds of dust and pungent diesel exhaust. It was nearly sunset by the time we arrived in the muddy village of Mawayani, the traditional starting place for Qollyur Riti pilgrims.

Although neither of us felt very strong, I decided that we'd better make the trek that evening. We hired horses to carry our backpacks, and we began the slow trudge uphill, moving with considerable effort up the stony path as the last sunlight moved off up one side of the valley – and a ghostly moonlight moved down the other to replace it. At intervals large crosses loomed up out of the darkness to the side of the path, with candles flickering around them and pilgrims, some of whom had walked for days from their villages to get this far, praying quietly. In smoky canvas stalls people were selling hot food, but exhaustion had overtaken hunger for us and we plodded onwards.

Higher up still, lines of people began to pass us. In the gloom, I couldn't see clearly what they were wearing, but it seemed to be some kind of fancy dress, with long capes and white balaclavas. Many of them talked in falsetto voices, giggling and teasing us as they passed. They overtook us easily, as Tim and I were getting slower and slower. But just as I was beginning to seriously regret having not spent more time getting used to the altitude in Cusco, we rounded a corner and Qollyur Riti opened up in front of us.

It was an incredible sight. The whole valley, headed by snowy peaks and glaciers that shimmered in the moonlight, was full of flickering lights and movement. It reminded me of a scene from *The Lord of the Rings*, where an enormous army masses outside the walls of a mountain fortress preparing for an epic battle. Acres of canvas shacks – interspersed with cooking fires and dark moving shapes – carpeted the hillsides, right up the steep sides of the mountains. In the middle of the valley was a low white church, the Sanctuary of Qollyur Riti (supposedly the site of a vision of Jesus on the cross), itself surrounded by even more pilgrims who surged in seemingly-endless processions around and in front of it. Several brass bands and the banging of drums echoed across the whole scene, which was frequently shot through by volleys of loud fireworks.

Neither of us got any sleep that night. Outside a frost was forming, the temperature soon plunging well below freezing. Tim threw up violently in the early hours, and every time I closed my eyes I felt either nauseous or about to suffocate or both.

I also spent the night worrying about how to find the right people to talk to. Most of the pilgrims, being indigenous *campesinos*, would speak the old Inca language

Quechua, and only a little Spanish at best. I also needed to find someone who had been coming to the valley for long enough to have noticed changes in the sacred glaciers. I suspected people might not be too friendly towards a foreigner wandering around asking unusual questions. And who were those strangely-dressed people we'd passed on the path? Many of them had been carrying whips, and I didn't fancy getting whipped for being too nosy.

I got up at sunrise to try and find water, climbing unsteadily up the mountainside to where a trickle of melt-water was emerging from underneath a thick bed of night ice. Down below, the festival was still in full swing: a loud-speaker had started up, and a preacher was haranguing the penitents gathered in their thousands outside the white sanctuary building. As I filled my plastic bottle with water, a stocky young *campesino* appeared on the slopes above, gathering dry brushwood for a fire.

'*Buenos dias,*' I croaked.

He grinned and squatted down next to me, telling me that his name was Francisco, and that he came from Marcapata, a village far away on the edges of the high Amazonian jungle. 'Is it your first time here?'

'Not in Peru,' I answered, admiring his colourful knitted woollen hat, 'but the first time ever in Qollyur Riti.'

'I've been coming every year for twenty years with my family,' he boasted.

'*Excelente.* Can you show me round later?'

He nodded. I'd found my guide.

*   *   *

Back at the tent, Tim was still out of action, so after leaving him with some water I packed a couple of chocolate bars and some spare clothes and set off slowly up the path. In the daylight the mountains which reared up in front were even more impressive – their snow-clad peaks glimmered as a morning heat haze rose from the rocky valley floor. Dozens of other people were also making the trek – some families with children, others in larger groups. We paused halfway, for me to catch my breath, and for Francisco to tell me something.

'Well,' he began, a little hesitantly, as he looked back down the valley, where the lower peaks on each side were blackened. 'Those were all *nevados* [snowy mountains] before, when I first came. Now there's nothing – just rocks.' I peered through the shimmering heat. It seemed difficult to imagine that those bare mountains, sun-baked and in places streaked with orange sand, had ever carried snow. Francisco went on, gaining confidence in his tour-guide role: 'When the snow and ice go away, there is no water to keep the plants growing – that's why it's not green up there any more. There will also be less for pasture and watering crops lower down. That's why it's bad to lose this snow.'

I asked him why he thought the ice was melting. He shrugged. 'Because of the sun. Everything's getting hotter.'

We turned and continued on up, sharing a bar of tasteless, chewy Peruvian chocolate. Further on the main valley split into three, and we continued up the middle path between two long ridges of moraine. The amount of vegetation on glacial moraine is often a good indication of its age, and therefore how long ago the ground had been covered by ice. This stuff, gravelly and full of odd boulders, was sparsely covered with a few dry bushes and tufts of grass. All around the path people had built little houses out of rocks and stones – offerings, according to Francisco, to the Apu mountain gods which would bring the offerer luck in his own home, and prevent the infiltration of evil spirits.

About a quarter of an hour further on, he stopped again.

'This is the place where I remember it was twenty years ago,' he told me after a pause, sweeping his arm around. 'Everywhere round here was under the ice when I first came here.'

I counted my paces up to the edge of the glacier ice, about 200 metres further up. If Francisco's memory was accurate, that meant the main Qollyur Riti glacier was retreating at about ten metres a year – a rapid rate by global standards, but far from unprecedented within the Peruvian Andes, where far quicker retreats have been scientifically measured. But the problem here was that the peaks themselves were not, by Peruvian standards anyway, very high: probably only about 5500 metres. And with the rate of retreat likely to accelerate as temperatures rise further, these glaciers – sacred or not – would not last long.

At the glacier edge itself there was a small crowd of people waiting to climb up the notches someone had cut in the ice – something that was much easier for Francisco with his sandals made from old car tyres than for me in

my big walking boots. On top of the glacier the solid ice
was covered by a layer of snow, but it was slushy under-
foot and the water ran in small trickles into cracks down
below.

Tramping steadily but slowly upwards amongst hundreds
of other footprints we reached a large cross, planted firmly
in the snow and guarded by a group of men – just like those
I'd seen the night before – carrying whips and wearing
shaggy costumes alternating red, yellow and black. These
were *pablos*, explained Francisco quietly, a kind of informal
police force whose job it was to make sure that no one got
into trouble or drank alcohol and that all the ceremonies
were performed correctly. Some had stayed here all night
despite the cold (it wasn't unknown for people to freeze to
death whilst discharging their *pablo* duties), and a big procession of them would come up in the early hours of the
next morning to bring the cross back down to the Qollyur
Riti sanctuary for mass. I caught the eye of one and he
approached me quickly, brandishing his whip.

'No hats,' he ordered. The hat was off and clutched
humbly by my side in less than a second.

We moved on. Higher up, people were busy enjoying
themselves, laughing and pushing each other down snow
slides. Many were dressed in traditional costume – one
group of women, wearing black skirts, wide-brimmed
hats and multi-coloured tops, threw snow at each other. On
one side three men were praying quietly, kneeling down in
the snow and muttering incantations. The view to the east
was staggering: Nevado Ausangate had appeared again

in glorious close-up, and even from twenty kilometres away I could make out almost every notch on the ice ridges that led up to its striking pyramidal summit. It was surrounded by similarly steep-sided mountains, one of which was entirely encased in white like an ice-cream. In the further distance were piled up row upon row of other unglaciated mountain ranges towards Cusco and beyond.

I began to watch where I was walking – some enormous blue crevasses were opening up on either side of the path. I realised with a shock that many of the smooth dips people were sliding into were in fact snow-covered crevasses. Francisco too was getting more cautious. 'People die up here every year,' he said. 'Maybe we shouldn't go much further.'

I asked him about the Apus, but he seemed reluctant and a little evasive – perhaps these were secrets that shouldn't be too casually shared with strangers.

'Everybody believes in them of course,' he said when I pressed him. 'They have a lot of power.'

'What kind of power?'

'Power over people's lives. Whether you can eat and live in security.'

'But if the Apus live in the snow, and the snow is disappearing, what does that mean?'

A shadow of doubt fell across his face. 'It means the Apus are leaving. I think bad things will happen to our people.'

He didn't want to say more. The sun was now high in the sky, its rays blindingly powerful on the ice. Francisco turned around, and together we headed back down to the camp.

I tried to doze in the tent for a couple of hours, and Tim turned up a little later carrying four kebabs and some fried maize-meal dough.

'I've been dancing,' he announced cheerfully, somewhat restored to his former enthusiastic self. 'I think people were impressed. They kept asking me what I was drinking and if they could have some.'

We sat cross-legged in the tent to eat. The meat was stringy, highly spiced, and Tim had no idea what animal it had come from.

'It smelt nice enough, anyway,' he said over a mouthful.

That night was even worse than the previous one. The noise was just as loud as ever, and I nearly threw up every time I closed my eyes. It was like being horribly, sickeningly drunk, with the world spinning round me. I lay there gasping. After two hours of fighting it, I gave up, pulled on my boots, put on my padded down coat and crunched off again across the frosty ground.

It was immediately clear that something was up – hardly anybody else seemed to be sleeping, and big groups of thirty or so were beginning to coalesce around strategically-placed embroidered pennants. The full moon bathed the whole scene in a soft, milky glow. Moving lights were visible up on the middle glacier already, columns of a silent army on the march. I remembered what Francisco had said about the *pablos* going up there in the middle of the night, and suddenly realised what was going on. The final pilgrimage had begun. And despite not having slept for the last three nights I had little choice but to follow.

For the first half hour I was alone on the narrow rocky path back up to the moraines. One group of villagers had been in front of me for a while, but I couldn't keep up with them. Their footsteps quickly faded into the distance, and I sat down on my own next to the path to get my breath back.

Far above me a long line of *pablos* walked in single file along the sharp crest of a moraine ridge, their flowing cloaks silhouetted like wizards' robes against the moon. There were also lights visible on one of the glaciers to the left, but in the semi-darkness I couldn't tell which path I was on or where I might end up. The scene was so unreal, and my exhaustion so complete, that it was like walking through the hazy landscape of a dream.

After a few minutes' rest, I pressed on up the steep slope, the narrowing path faint but still just about distinguishable in the dim light. I had paused again to catch my breath when one of the large rocks by the side of the path began to move. With a shock I noticed that I was surrounded by a large group of stationary *pablos* – whips, balaclavas and all.

'Acomayo!' someone called out.

I gulped and said nothing, unsure how to reply.

'Acomayo!' several others piped up, some in ringing falsettos that echoed shrilly around the rocky gullies above.

'Er, no, I'm from England,' I announced in Spanish, my

admission sounding deeply foolish in the silence. I seemed to be talking to several dozen people at least, and added hopefully: 'But I can be from Acomayo if you want . . .'

There was laughter, and two or three people stood up to greet me.

'Sit down with us,' one of them said. 'What's your name?'

'Mark. M.A.R.K.' I spelled it out.

'Marsh. Mart.' They practised saying it. 'Then you can be Pablo Mark from Acomayo,' the friendly one said, then added sternly: 'You should join us because you have to be safe up here.' I was only too happy to comply.

'Do you have any whisky?' someone else broke in. Obviously the alcohol prohibition wasn't universally applied.

I didn't have whisky, but I offered water and chocolate, both of which were gratefully accepted. In return I was given a handful of coca leaves, which I crunched up and tried to stick in my cheek like everyone else. But the dryness of the leaves almost made me retch, as did the unpleasant taste – stale and bitter, like rancid tobacco. But the juices slowly produced a gentle numbness in my mouth, and the cold and tiredness began to lift a little. I understood then why coca leaves retain such a vital role in traditional *campesino* society. Chew coca and your hunger and thirst don't matter as much, even if you still have nothing to eat or drink.

'*Vamos,*' someone said, and we all stood up, before moving off in single file amidst a chorus of 'Acomayos' and a lot of jostling for position. Everyone referred to everyone else by the title *pablo*.

'Pablo Mark, *como estas?*' said a voice behind me every now and then.

I stayed in line, taking one step at a time, and somehow gathered strength and warmth from having been adopted into a community. We reached the edge of the glacier – further on from the one I'd visited during the day – at about 4 a.m., and stopped in a tight huddle on the rocks. It was now the coldest part of the night, and I was freezing even with someone else's poncho on my knees and people hunched against me on all sides. Everyone else seemed immune to it, swapping banter and the occasional alcoholic drink. Someone even fell asleep, snoring loudly amongst the seated throng.

Dawn began to break at around 5.30, sparking a sudden rush to get onto the glacier before the sun came up. Together we formed a single file, grasping a line of knotted-together whips which served as an impromptu rope. A few more steps and we were up on the ice itself.

Everyone spread out around the big wooden cross, some people cutting squares in the snow-crust to make into little houses, and placing flickering candles and coca leaves inside them as offerings to the Apus. To the east the high slopes of Nevado Ausangate were already catching the first rays of the sun, and as the bright light began to hit us, the snow crystals all around glittered and sparkled like gems.

Already people from the other glaciers were returning to the camp, streaming back down the valley in long, inexhaustible lines. For a while I just sat there, whilst the sun – worshipped by the Incas as the most powerful god of all – grew steadily in strength and power, and the icy Apus prepared to continue their retreat up towards the peaks from which there would be no return.

JACABAMBA

Most people probably wouldn't readily associate the word 'tropical' with the word 'glacier'. Yet Peru's Cordillera Blanca tells a different story. Located only about nine degrees south of the equator, it runs north to south for nearly 200 kilometres in a chain of immense summits, twenty-seven reaching over 6000 metres in height. Its highest peak, the massive 6768-metre Nevado Huascaran, is not only Peru's tallest mountain, but also of anywhere else in the entire tropics – as well as being higher than any in North America, Africa or Oceania. In fact the Andean range as a whole (of which the Cordillera Blanca is a part) is second only to the Himalayas both in scale and altitude.

As a consequence, within the last two decades, the Cordillera Blanca – and its principal town, Huaraz – has become a mecca for climbers from all over the world, eager to test themselves on its precipitous ice-clad summits. Its main streets are now full of climbing shops, renting everything from crampons to tents, jostling for space with pizza restaurants and chicken fast-food outlets. And like an Alpine resort, the town itself is dominated by glistening snowy peaks. The towering bulk of Nevado Huascaran itself is visible from the town's main square.

Not surprisingly, since it serves a region containing a quarter of the world's tropical glaciers, Huaraz has also attracted its fair share of glaciologists. There were several already in residence, from as far afield as Austria, France and Scotland, when Tim and I arrived at 'Mi Casa', a small hostel near the centre of town which is in fact run by the Peruvian glaciologist Alcides Ames. Now greying and retired, Ames almost single-handedly pioneered the science in Peru, making the first studies of how different glaciers throughout the Cordillera were responding to climate change.

The main subject of his investigations was glacial mass balance: whether the ice fields were losing or gaining size. Peruvian glaciers are particularly sensitive indicators of climate change because their location means that melting continues all year round – unlike higher-latitude glaciers such as those in the European Alps, which get a respite from melting (and a dose of new snow) during the winter months. Peru is close enough to the equator for there to be only small differences between winter and summer, so temperature rises which only have a summer effect on temperate glaciers act year-round in the tropics, doubling their impact.[1]

Just being back in the Cordillera Blanca left me abuzz with excitement and recognition. I could happily spend years sitting staring up at the endlessly-changing mountain vistas, lit up one minute by a brief shower of hail, and the next by a delicate orange sunset. During our first couple of days, I kept sneaking up the stairs of the hostel every five minutes to look at Huascaran, often wreathed in cloud and rising up a good thirty miles down the broad Callejon de Huaylas valley.

I wanted to hire a car to revisit the spectacular Laguna Llaca, a tiny dammed lake under the near-vertical face of 5800-metre-high Nevado Ocshapalca – I had a photo of myself aged six, wearing a red jumper and a straw sombrero, posing on the dam wall. But Alcides informed me with regret that the road was no longer passable.

So Tim and I contented ourselves with checking through the equipment, under the watchful eye of Alcides, who insisted we try out the primus stove he was lending us (resulting at one point in foot-high yellow flames and vast quantities of black smoke), and even made us pitch the tent in his shady courtyard. I toured the nearby indoor market, purchasing packet soups, quick-cook pasta, onions, tomato paste, large quantities of dried fruit and two jars of jam. It was a hive of activity, with chickens (both alive and dead) everywhere, goats and all kinds of fresh vegetables. On the street outside old women were selling bright green alfalfa, grown on the irrigated slopes high above Huaraz and brought down in huge bundles on the backs of donkeys to sell as livestock forage.

I asked Alcides if he'd ever been to the Jacabamba area, where my father's glacier photos had been taken. He examined them carefully, as we shared a breakfast of scrambled eggs and coca tea.

'Well, I know the mountains on this side, but Jacabamba is on the eastern side of the Cordillera Blanca. Hardly anyone goes up there, and I've never been further than Huari.' With no tourist infrastructure, we'd have to look after ourselves. Even Huari, a much smaller town than Huaraz, was seldom visited by foreigners.

We got onto the bus just after midday, watching from inside with increasing admiration as cages of chickens, sacks of potatoes, and even two large iron bedsteads were loaded on the roof above us. The first part of the journey took us past a very pretty lake I remembered clearly – although not fondly. We had camped there one night after having travelled up from Lima, and my whole family had been utterly incapacitated by altitude sickness. Instead of admiring the view, we'd spent the entire night passing the sick bowl between every single family member except my father: whose job was to trudge outside, empty it and then rinse it in the lake before bringing it back to the tent to be refilled. It later turned out (according to him, anyway) that he'd been misled by an incorrect roadsign, and we were five hundred metres higher than he'd originally intended. 'It was the worst night of my life,' my mother remembered years later. 'I really wanted to die . . . It was even worse than giving birth!'

This time round, Tim and I both felt fine – give or take a little shortness of breath. The unpaved road led through a dingy tunnel at the top of the pass, and then down through a series of spectacular hairpin bends, which the driver took at terrifying speed. No one else on the bus looked the slightest bit bothered, so I sat there stoically as the bus lurched repeatedly and sickeningly close to the edge. It took over six hours to reach Huari, and we arrived in the dark – lugging all our gear through the deserted streets to a half-built hostel.

Our first mission the following morning was to find two donkeys. My father's expedition had required a dozen horses and at least as many porters: but then they had also been carrying heavy drilling equipment, mapping gear, and enough food for three weeks. Now there was just the two of us – but with all our climbing gear and food we still had too much weight to carry by ourselves. Usually there are donkeys all over the place in Andean towns, but for some reason Huari that morning was donkey-free.

After an hour of fruitless wandering and questioning, we almost fell over ourselves in delight at the sight of one scrawny old nag plodding miserably up a narrow back-steet, straining under a fifty-kilo sack of fertiliser. (I've always thought that if you're unfortunate enough ever to be reincarnated as a Peruvian donkey, the only thing to do is grin, bear it, and hope to die again as soon as possible.) Its owner, a similarly scrawny old man in brown trousers and a thickly knitted blue jumper, seemed amenable, so we accompanied him up a cactus-lined path to his thatched adobe house high on the slopes above the town.

It was an idyllic spot: sheaves of yellow and purple maize hung drying from the wooden beams outside, and succulent-looking melons were draped over a fence in the shade of several large trees. The old man's neighbour explained that all the donkeys were busy bringing in the maize harvest, so it would unfortunately be impossible for us to hire one in Huari – though we could try the next village ten miles further on. This sounded suspiciously like the operation of the iron laws of supply and demand within the universal principles of the free market, and sure enough a suitably high price was sufficient to lure two donkeys – and boys to drive them – away from their work in the fields for the day.

We set off straight away, plodding up the stony path behind the two donkeys and their drivers, lathering ourselves with sun cream. I was entranced afresh by the patchwork beauty of the Andean landscape. Tiny fields, their colours varying from the deep green of barley through the light brown of newly-turned earth to the burnt yellow of maize, fitted neatly together across the steep slopes, interspersed by walls of prickly-pear cacti and tall eucalyptus groves, whose leaves pattered in the breeze like gentle rain. All around gurgled little irrigation streams, the fresh water collected from glacial runoff high above and transported through ancient networks of ditches, many dating from Inca times, down to the intensively-cultivated lower slopes. The unworked land was parched and cracked by the dry season sun, but where water flowed life abounded – with lush grass, brightly flowering shrubs and a never-ending chorus of birdsong.

A little further on, the first snowcapped mountains – our eventual destination – came into view, rising precipitously into the clear blue sky at the head of the valley. The boys and donkeys had long since disappeared in front, but there was only one path, which climbed gently up through boulder-strewn meadows and occasional tangled patches of forest. We found the donkeys again in the late afternoon, but the boys were nowhere to be seen, so we left them a note and a couple of dollars and shouldered our unloaded gear for the last two miles.

I had a particular camping spot in mind, described by my father in his diary as 'a rocky hollow near a little waterfall' – and some of the grassy humps next to a little gorge near the head of the valley looked perfect. We set up camp in quickly-fading light, and by the time I'd finished cooking our first meal, a fantastic display of stars was already arching across the frosty night sky. For half an hour we lay with our heads out of the tent, watching shooting stars, before the cold drove us inside to sleep.

We spent the next day acclimatising – trekking up through the coarse grass on the north side of Jacabamba valley to a spectacular vantage point on the narrow ridge which crested it. With the easy confidence born of several years' climbing experience, Tim bounded energetically about, but I found myself gripping the rock tightly and rather nervously as we traversed it. One slip, I kept telling myself, would be all that was needed to bring the trip to a painful and disappointingly inconclusive end. Perched on a ledge of lichen-covered rock which I firmly declared to be our high point for the day, I scanned the mountains in front, searching for locations my father had described. Towering over the whole scene stood Nevado Rurichinchey – a 6200-metre pure white pyramid, its icy ridges channelled with hundreds of little avalanche gullies.

One black arc of rock stood out from within the tumbling ice of the nearer glacier – Nunatak Ridge, named by my father (almost certainly the first person ever to set foot there) using the Eskimo word for 'a mountain completely surrounded by ice'. At least I thought it must be Nunatak Ridge: I knew from his diary that it had taken my father an hour to pick his way through the crevasses from the glacier edge to this island of rock in the middle of its icy sea. In the process, he narrowly missed being swept away in an enormous avalanche, which threw huge ice boulders across the route he had passed only moments before.

Yet now this ridge had only a tiny finger of grey ice protruding beneath it. Had the glacier really retreated that far? If so, this ridge would soon need renaming, its icy sea having disappeared. It would also indicate a rate of glacial retreat much faster than I had expected. But without a clear photo to compare it with, I couldn't be sure.

I did have a clear photo my father had taken from the ridge opposite: a towering arête leading up to the highest summits, with orange scree coursing down its lower slopes and glacier ice above. Tim suggested we aim to bivouac up on the snow itself, reducing our climbing weight by leaving the tent and most of the other gear in the valley. We set off early the following morning, following a weedy path that zigzagged up the first section of the slope. Here I was following in my father's exact footsteps – this path had been hacked out of the hill by the porters of his expedition in order to get equipment and pack animals to a higher camp. The campsite itself bore no trace of human habitation, but with several grassy mounds and a small waterfall at the back it must have made a pleasant base.

We continued on up, labouring even under our lighter packs with the altitude, scrambling up the scree and slopes of solid rock. Occasionally a stone bounced past us from somewhere high above, and unseen avalanches rumbled out of sight, their vibrations shaking the valley. The view from the top of the ridge was stunning: glaciers were below us on both sides, and in the distance successive black mountain ranges led east to the great basin of the Amazonian jungle.

We set up camp straightaway on a smooth area of untouched snow on the far side of the ridge; Tim boiling some soup on the cooking stove whilst I used my ice axe to dig out a flat patch of snow for us to sleep on. We were almost entirely surrounded by a circuit of bright white mountains and, as the sun dropped behind the precipitous slopes of Nevado Rurichinchey, bitter cold descended.

At an altitude of over 5000 metres it was virtually impossible to sleep, and I was happy to get up again at first light, brushing the frost from my sleeping bag and struggling to put on my frozen boots. Looking east, into the rising sun, was the most spectacular mountain view I've ever seen. A sea of cloud stretched into the far distance, with only the highest snow-covered peaks – including the one we were camped on – poking out above. It was as if the rest of Peru had disappeared. A hundred kilometres south the Cordillera Huayhuash, Peru's second highest mountain chain, floated above the mist like mysterious inaccessible peaks in a Chinese watercolour.

Climbing the ridge was surprisingly difficult: the frost-shattered red and brown rocks were unstable and easily pulled out of place, providing bad handholds. At one point the ridge was intersected by a steep gully with near-vertical drops on each side. One slip here and the fall would probably be a fatal one. Tim set up a belay anchor and paid the rope out as I inched across. Then we were onto pristine snow, our crampons providing superb grip on slopes of sixty degrees or more. I kept a watchful eye out for buried crevasses, probing the snow in front with my ice axe. Back on the ridge the snow had formed into a dangerous overhanging cornice, and I edged away from the blue line which showed a large area was set to topple off.

I was keeping my father's photo close to hand and the background landscapes were finally beginning to line up with my printed picture. It was Tim who discovered the exact spot, with even the same triangular rock in the near foreground.

Glancing at the view and then back at my photo, I could see that a dramatic change had indeed taken place. More than half a kilometre had disappeared from the main glacier at the head of the valley. Even the heavily-crevassed icefall in the foreground of my father's picture had wasted away, leaving just a few isolated patches of thawing snow. And the melting was clearly continuing apace: several new streams and waterfalls had sprouted along the edge of the retreating ice.

I was now more than halfway to answering my father's question about the fan-shaped glacier above the lake. One thing was certain: it wouldn't look the same as it had twenty years previously. If it had been affected in the same way as the larger glacier in front of me, it would probably have shrunk by about half. But since the slopes which held it looked from the photo to be almost sheer, perhaps it would have remained proportionately less affected than this flatter glacier, which would have more mass to lose from the same rise in freezing level. There was only one way to find out, and we descended back to our base camp in a sudden thunderstorm, ready for a trek up to the lake the following day.

I hardly slept again: not just because of the altitude this time, but because I was itching with curiosity. The whole of my journey had begun with that question two years before, and although I hadn't yet seen it, I knew that the changes I would witness on this glacier symbolised the wider climate changes taking place right across the planet.

The route was pretty straightforward, over mounds of old moraine and across rubble-strewn riverbeds, up the left-hand fork from the top of the main valley. There was little vegetation – just a few scrubby bushes on the slopes higher up, and some intrepid little blue flowers that had sprung up amongst the boulders. Nor was there any sign of a path on this side of the valley.

From time to time we stopped. I kept my father's (now rather grubby) photo handy, looking again at that enormous fan of ice, completely dominating the little iceberg-strewn lake. When we reached the lip of an old terminal moraine about three hours above our campsite, I stopped again, with a flutter of trepidation. I knew that the answer to my father's question lay just a few metres behind it.

Then I pressed on, over the crest and was looking straight at the scene my father had last witnessed over twenty years before.

For the first few seconds I couldn't speak. Then I shouted to Tim.

'It's gone!'

He hadn't even realised we were in the same place.

'What's gone?'

'The whole fucking glacier's gone. This is the same lake!'

I dropped my rucksack on the ground and shoved the old photo in his face.

'Bloody hell,' he said, studying it and then looking up at the current view. 'That's unbelievable.'

I could scarcely believe my eyes either. The lake was the same – if rather larger, probably because of the extra meltwater – and the rock walls encircling the whole arena just as sheer: but the mass of ice and snow which previously dominated the scene had melted completely away, leaving only a few large mounds of dark red grit, streaked white with recently-fallen avalanche debris. The main glacier was still high above, but it had also thinned enormously.

It took a while for the magnitude of our discovery to sink in. Tim took some new photos – from exactly the same vantage point as my father – whilst I sat on a patch of grass and tried to comprehend the scene. I had fully expected the glacier to be smaller, perhaps for it to have even retreated away from the lake . . . but its complete disappearance was a hammer blow.

I wondered how my father would react. I knew he'd been fond of this place – and it had indeed been spectacular. Looking back at the dog-eared photograph, it looked both awe-inspiring and invincible, and yet its existence had proven all too transitory, its might surprisingly frail. This glacier had probably survived since the end of the last Ice Age, only to vanish completely within the space of a single generation.

Back in Huaraz, Alcides Ames took the news with a grim shake of the head. It wasn't surprising: the same process was happening all over the Peruvian Andes, and in mountain ranges all over the world. One of the first glaciers he'd studied, as a young graduate back in 1968, was Glacier Broggi, in the northern part of the Cordillera Blanca. On a return visit in 1998, he'd been shocked to discover that Broggi was 'dying'. It had retreated over a kilometre since the 1930s, and by the mid-1990s was losing seventeen metres every single year – a rate of retreat that was still increasing.[2] He hadn't been back since, but 'even then there was hardly anything left. It's probably now disappeared completely,' he added sadly.

Similarly rapid rates of decline have been measured throughout the Cordillera Blanca. The glacier underneath the beautiful ice cone of Artesonraju has retreated by over a kilometre since the 1930s; its tongue is now 500 metres higher up than previously.[3] Glacier Yanamarey, also one of those studied by Alcides Ames, has lost half its surface area in the last thirty-five years alone.[4] In the whole of the Cordillera Blanca, glacial retreat has been speeding up dramatically since 1980 – and is now three times faster than before.[5]

Between 1970 and 1997 the Cordillera Blanca is estimated to have lost 15% of its glacier surface area: a total of 111 square kilometres,[6] or about a third the area of Britain's Isle of Wight. The rate of loss has been equally drastic in other Peruvian mountain ranges, many of which, because they started from a smaller base, have proportionately lost even more ice than the Cordillera Blanca. Three of the smaller ranges have lost nearly half their glacier area, whilst many of the eastern cordilleras have lost around a third.[7]

Although the Peruvian Andes account for the vast majority of the world's tropical glaciers, there are also smaller glaciated mountain ranges in East Africa and New Guinea – and they too have been seriously reduced by climate change. Mount Kenya's glaciers lost three-quarters of their entire extent during the twentieth century.[8] Kilimanjaro is losing its famous snows at an ever-accelerating rate: this huge volcano's glaciers declined by 80% between 1912 and 2000, the remaining ice fields fracturing into numerous smaller pieces.

When the American glaciologist Lonnie Thompson drilled a fifty-metre-deep ice core on the top of Kilimanjaro he found that organic matter trapped in the ice at the bottom of the core could be carbon-dated to 9000 years ago, indicating that these glaciers had survived all the

Ye Yinxin smiles for the camera, despite being the last person left in her entire village.

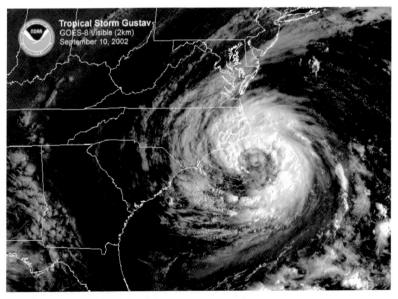

Satellite image of Tropical Storm Gustav, just when the storm was at its most intense on the Outer Banks. I was just north of Cape Hatteras at the time – in the worst of it.

The storm surge floods over Highway 12 on the Outer Banks, North Carolina, during Tropical Storm Gustav.

The author scribbling notes in the headlights of our broken-down bus, somewhere late at night on the road between Huari and Huaraz.

The coldest I have ever been whilst putting on a pair of socks, but also with the most dramatic backdrop. Sunrise at our high ridge camp above Jacabamba valley; 6200-metre Nevado Rurichinchey is the highest peak behind.

Looking down at our Jacabamba ridge camp from higher up the glacier, with the sea of cloud obscuring the eastern Andes and Amazonian basin behind.

My father's original 1980 photo of the fan-shaped glacier at the top of Jacabamba valley, in the eastern Cordillera Blanca, Peruvian Andes. It was this photo which convinced me to retrace his steps and triggered my journey around the world. Photo credit: Bryan Lynas

The same place at the head of Jacabamba valley where my father's glacier used to be. Note also the severe thinning of the icefield on the skyline above the lake. Photo credit: Tim Helweg-Larsen

Left Under the haze of woodsmoke and dozens of Peruvian flags, vast processions gather near the sanctuary in Qollyur Rit'i.

Below A group of pilgrims – including *pablos* in fancy dress, stand around a cross on the glacier above Qollyur Rit'i. Note the crevasse just behind!

Glacier Sullcon, the most important ice reservoir which supplies Lima with water via the River Rimac. The rubble ridge in front is, remarkably, the continental divide of South America.

Looking down the valley at where Glacier Sullcon used to be. The desert-like mounds are moraines, indicating that the ice retreated from there relatively recently. It was at about this place that the altitude sickness caught up with me – with devastating consequences.

natural climate changes that had swept the world since the end of the last Ice Age. Yet at current melting rates, every bit of ice on Kilimanjaro will have melted by 2015–2020: evidence suggesting that temperatures in the twenty-first century will be warmer than at any time for tens of thousands of years.[9] There are now less than three square kilometres of ice left on the West Papuan mountains in the deep interior of New Guinea island: down from a probable twenty square kilometres a century and a half ago.[10]

In fact, just about every glaciated mountain range on Earth is experiencing massive ice losses. Somewhere between 500 and 1000 cubic kilometres of water have melted from Chile's Patagonian ice caps in the last fifty years – enough to add a couple of millimetres to global sea levels.[11] In the United States, the original 150 glaciers within Glacier National Park are now down to a mere fifty – and most of these are tiny remnants of their former selves. Local geologists have calculated that the whole park lost 73% of its glacier cover between 1850 and 1993.[12]

In Alaska, meanwhile, researchers have used laser alti-metry from overflying aircraft to calculate glacier wastage, discovering that 95% of Alaskan glaciers are thinning – and that since the mid-1990s the thinning rate has doubled. The melting Alaskan glaciers are now adding as much to global sea level rise as the thawing Greenland ice sheet: around 0.2 millimetres a year.[13]

In the European Alps, about half of the total glacier mass is thought to have disappeared between 1850 and 1990.[14] Moreover, a study published in 2002 concluded that there has been a 'drastic acceleration of retreat since 1985'.[15] Extrapolating current trends into the future, the Alps are likely only to have a quarter of their original ice cover left by 2025, and only the highest and biggest glaciers will survive into the twenty-second century.[16] During the record-breaking hot summer of 2003 Swiss glaciologists found the glaciers melting at an exceptionally rapid rate – up to ten times faster than the previous average.[17]

Even the mighty Himalayas have been affected. The Rongbuk Glacier, which tumbles down from the north face of Everest, retreated between 170 and 270 metres between 1966 and 1997: a rate of retreat of around eight metres a year.[18] When Japanese scientists visited the glaciers in Nepal's Hidden Valley in 1994 – repeating work they had done two decades previously – they discovered that most of the glaciers now ended thirty to sixty metres higher up the mountain. In addition, the valley's longest glacier, the Rikha Samba, had not only retreated by 200 metres in length, but was still melting unusually quickly.[19]

In total, the world's mountain glaciers are now thought to be losing about a hundred cubic kilometres of water every single year – more than the entire volume of Lake Geneva in Switzerland.[20] The worldwide thaw began to accelerate in 1977, and since the end of the 1980s – following a succession of the warmest years ever recorded – the rate of global glacier melt has speeded up even further.[21] The only glaciers currently gaining mass are those in wet maritime areas like southern Norway, where melting has been offset by an increased snowfall.

Every glacier has what is known as an 'equilibrium-line altitude' – the line of balance above which accumulation of new snow outstrips the rate of melting, and below which the glacier is permanently losing volume to evaporation and water runoff. One recent estimate has put the world-wide increase in the equilibrium-line altitude at around 200 metres between 1960 and 1998[22] – an increase which can only have come about because of rising global temperatures.

This was recently backed up by two of Switzerland's most prominent glaciologists, Wilfried Haeberli and Martin Hoelzle, who pointed to the discovery of the Oetztal ice man (whose 5300-year-old frozen body thawed out of an Austrian glacier in 1991) as evidence that Alpine temperatures were reaching unprecedented heights.

'Recent progress in analysing worldwide glacier mass balance data has made it possible to state with reasonable certainty that average glacier mass loss now roughly corresponds to overall effects from estimated human-induced greenhouse forcing,' they wrote recently, adding that with the current acceleration trend the world was now moving beyond the limits of post-Ice-Age natural variability.[23]

By coincidence, an international conference on mountain ecosystems was also taking place[24] in the week that I was in Huaraz. I sat scribbling notes whilst speaker after speaker took the podium to reveal their latest research on glacial retreat. The Austrian glaciologist Georg Kaser spoke of how glaciers were like a bank account: 'income' was snow accumulation, whilst 'withdrawals' were meltwater runoff. So long as this account stayed balanced, meltwater flowing into the rivers in the dry season would be balanced by new snow falling in the wet season, keeping the rivers running all year round.

However, the system had recently tilted out of balance: more of the glaciers were melting each year than were being formed anew by snowfall. Right now, he pointed out bluntly, 'we are living off our savings'. Each retreating glacier was releasing more water every year, increasing the amount of water available in rivers for drinking, agricultural and industrial use. But this wouldn't last. When the glaciers disappeared completely, so would this extra water – and the year-round mountain rivers most of Peru's desert population depends on for survival would become seasonal.

Next to speak was the French hydrologist Bernard Pouyaud, who had studied one of the glaciated side-valleys in the Cordillera Blanca to see if the amount of water in the stream equalled the amount of rain and snow falling during the year. It didn't. 'The annual precipitation was 800 millimetres, but the runoff was double – so we could only conclude that the rest must come from glacial melting.' No one could say how long this extra runoff would continue, Pouyaud went on, but his guess was that 'in thirty years there will be a big decrease'.

Benjamin Morales Arnao, president of the Andean Institute for Glaciology and the Environment,[25] used his talk to give a stark warning via the assembled media to the residents of Peru's capital Lima. 'Glaciers in the lower cordilleras, like the Cordillera Central above Lima, will disappear. Much of the water coming into the Rio Rimac which flows to Lima is glacial and, if those glaciers disappear, it will be a huge problem for the city.'

I was due to leave for Lima the very next day. But Morales Arnao had given me an idea: to make an excursion up to the Cordillera Central to see for myself the state of the glaciers that keep the country's capital city supplied with water. However, Tim was already booked on a flight back to England – so I would be making the trip back up into the high mountains on my own. I knew this would increase the dangers considerably – if I had an accident or got struck down with altitude sickness there would be no one to help. But I calculated that I was now well enough used to the altitude to make a solo trip viable.

Unfortunately, I was wrong.

LIMA

Not for nothing are mountains known as 'nature's water towers'. All the major rivers in the world – from the Nile to the Rio Grande – have their headwaters in mountain ranges. It has been estimated by the UN that half the world's people rely on mountain-generated water to grow food, generate electricity, sustain industries, and, most importantly of all, to drink.

In humid parts of the world, mountains supply 30 to 60% of downstream fresh water (as opposed to water supplied by rain falling on the lower reaches of a river's catchment), whereas in semi-arid or arid areas the proportion is

70–95%.[26] On Peru's Pacific coast, it's 100%. This is hardly surprising: the country's coastal strip is part of the Atacama desert, which runs south from the Ecuadorian border to central Chile and is one of the driest places in the world. No rain at all falls most years. Yet this region is the source of half of Peru's national agricultural production – from sugar cane to lemons – in fifty-two irrigated alluvial valleys coming down from the Andean mountains.[27]

One of the most productive of all is the Santa valley, whose river comes down from the Cordillera Blanca and whose water is used intensively to generate electricity and supply human populations, as well as for cultivation. After studying forty years of hydrological data from the valley, a team led by the Austrian glaciologist Georg Kaser found that 'during the dry season the runoff is almost exclusively due to the glacier melt', water which 'is of essential importance for the highly populated and cultivated valley . . . particularly during the dry season'. However, Kaser concluded ominously, if the current rate of glacial retreat continues, 'glaciers will strongly shrink or even vanish' from some catchment areas – so their vital contribution to dry season runoff 'cannot be expected [to continue] long term'.[28] The situation is the same for all of Peru's irrigated coastal valleys: no glaciers will mean no runoff in the dry season.

Peru's capital Lima is built in one of these valleys. This enormous metropolis is home to nearly eight million people, and is the largest desert city in the world after Cairo. But apart from a few rapidly-diminishing wells, every drop of water supplied to Lima flows down the valley of the River Rimac from the snowcapped mountains of the Cordillera Central.

As a result, the twenty-first century will see the collision of two opposing trends: on the one hand, Lima's population is likely to grow to about 10 million people by 2015 (the current annual rate of growth is 200,000), all of whom will need fresh water to survive.[29] On the other, glaciers in the Cordillera Central – which have already reduced in volume by a third between 1970 and 1997[30] – will disappear altogether in just a few decades unless global temperatures stop rising.[31] Consequently, the Rimac River, which through the late twentieth century has been temporarily charged with additional meltwater from the rapidly-retreating ice fields, will suddenly – and disastrously – dry up for half the year.

It's difficult to imagine quite how a massive Third World city might cope with a crisis on this scale. With no water supply for six months every year, life will quickly become impossible. Where will its residents go? There is no spare land in the mountains, and few could survive in the jungle. Whilst the rich could pay for fresh water to be trucked in, the poor – the massive majority of Lima's population, who already have difficulty accessing reliable water supplies – will be forced to move or die. And Lima is not the only city to be affected by this problem. Many of Peru's other major cities, like Arequipa in the south and Chimbote in the north, face a similar plight.

Nor is the looming crisis confined to Peru. Bolivia's capital La Paz and Ecuador's capital Quito both depend on glacial runoff. One Bolivian glacier, located only twenty kilometres from La Paz and a key water source for the highland city, is likely to disappear within ten to fifteen years; the stream it currently feeds will then be entirely dependent on regular rainfall to keep flowing.[32]

Warnings have also been sounded in Asia, where half a billion people in the Indian subcontinent who depend on rivers flowing down from the Himalayas could be left without water – within this century – if the Himalayan glaciers melt. Half India's hydro-electric power is currently generated by glacial runoff – and scientists have calculated that an eighth of that is temporarily increased meltwater from glacial recession.[33]

The River Indus – the only water supply for millions of people on the drought-stricken lower plains of Pakistan – gets 90% of its lowland flow from the mountains of the Karakoram and Western Himalaya.[34] Yet recent studies of satellite pictures of the headwaters of the River Satluj, a tributary of the Indus, showed that all eight major glaciers were in retreat, each having lost between one hundred and a thousand metres in length in the last forty years.[35]

Researchers co-ordinating a worldwide programme of glacier monitoring from satellites (the Global Land Ice Measurements from Space, or GLIMS programme) have found that, despite their immense height, the Himalayas are actually one of the worst-affected mountain ranges in the world. As US scientist Jeff Kargel, GLIMS' international co-ordinator, told reporters at a conference of the American Geophysical Union: 'Glaciers in the Himalaya are wasting at alarming and accelerating rates.' He pointed in particular to the Gangotri glacier between Kashmir and Nepal, which lost nearly a kilometre and a half during the twentieth century. This glacier feeds the Ganges river basin, which in turn supplies water to hundreds of millions of people downstream – including those living in New Delhi and Calcutta.[36]

As with Lima, if the worst happens and many of the crucial Himalayan glaciers disappear, hundreds of millions of people will be faced with moving or dying of thirst. The scale of this threat is so colossal that it almost defies comprehension.

Arid Central Asian countries like Uzbekistan and Kazakhstan, which depend on glacier runoff from the Tien Shan mountains on the western edge of the Tibetan Plateau, could be even worse hit. As Kargel himself says: 'It's fair to say that Uzbekistan's economy and urban zones will either suffer devastating collapse in the future, or the people will have to adopt different ways of living on little water. Many governments and several nations may rise or fall with the loss of these glaciers, depending on how they deal with the issue.'[37]

In order to find out how Peru's water authorities were planning to cope with this threat, I arranged a meeting with Joel Campos, a senior manager at SEDAPAL, Lima's water utility.

Although surrounded by shanty towns, the SEDAPAL complex looked both well-resourced and efficient. Briefly lost after being deposited in the wrong place by a taxi, I gave myself an impromptu tour of the water purification facilities – including an enormous warehouse-sized tank of crystal clear water, surrounded by a thicket of pipes and sluice gates. The office was a space-age block of round towers and high passageways, all faced with shiny blue glass and set amongst enormous green lawns. A mysterious artificial waterfall appeared and then disappeared behind it, tumbling down from the bare red rock of the desert foothills.

Joel Campos was stout and cheerful, and after the introduction formalities were over, I explained my concerns.

'We have enough water now, but we're going to have problems by 2005,' he declared frankly, spreading out several large maps on a table. 'This map is of the upper catchment of the River Rimac, where all Lima's water comes from. You can see here in the higher areas,' and he indicated the mountain tops of the Cordillera Central, 'there are blue contours where you find ice and snow?' I nodded. 'Well, all of this used to be blue contours too.' He waved a hand over a large highland area where brown lines indicated bare rock. 'All this area used to be snow and ice. When you go up there now you can see clearly where the snow was – it's all empty, desert-like and nothing grows. And already we can see that the amount of water coming from the mountains is beginning to reduce because of the disappearing ice.'

SEDAPAL was well aware of the threat posed by melting glaciers, Campos insisted. For that reason a massive programme of construction works was envisaged – aimed both at capturing runoff in dams and at piping water right through the mountains from lakes on the other side of the continental divide. But dam-building, as well as being expensive, had its own obvious dangers in a high-risk earthquake zone like Peru.

But, despite the drawbacks, some dams have been built, including a big new lake at a place called Yuracmayo. The Rio Rimac's flow has been augmented artificially by the recently-constructed Transandine tunnel, which brings water from a neighbouring catchment, and another tunnel is currently planned which would increase this extra input still further. But this second tunnel comes with a $120 million price tag, and the deeply-indebted Peruvian government has shown no sign of stumping up the cash.

'We have to contend with the political effect of not having any money, as well as the greenhouse effect,' Campos complained. 'Everything is ready for the second tunnel, but nothing is happening.'

I asked what else was planned.

He shrugged. 'We're exploring all the options – even the desalination of sea water with petrol generators.'

'Wouldn't that worsen the greenhouse effect?' I asked, pointing out that petrol generators would produce lots of the greenhouse gas carbon dioxide.

'Sure,' he replied. 'But what else are we going to do? Anyway, it's probably too expensive.'

With the maps spread out in front of us, I realised that this was a unique opportunity to finalise my plan of visiting the most important glacier in the Rimac basin: Glacier

Sullcon. This was being studied by one of the glaciologists I'd met back in Huaraz who had showed me grainy pictures of a fat tongue of ice surrounded by sharp peaks.

'Can I get to Sullcon?' I asked Campos.

'Of course,' he replied, tracing a route on the map. 'There's a SEDAPAL road along here right the way to the Yuracmayo lake – that's one of the dams we built. From there you can walk seven kilometres to the Sullcon glacier. No problem.'

'Can I take a taxi from Lima?'

He laughed. 'I don't think so – the road is in bad condition. You'll need a four-wheel drive.'

He gave me the map as a farewell present. 'You can't buy this anywhere, so it might be useful.'

I objected politely.

'No, take it, take it. You'll need it,' he insisted.

I spent the next day rushing around different parts of Lima with one of the hotel's regular taxi drivers, a sharp-witted young man called Maycoln, who was itching to ditch his battered old taxi for a powerful four-wheel-drive jeep.

It arrived first thing the next morning, and we admired it together: a white Toyota Hilux, with good ground clearance and new, well-treaded tyres. It was a typical Lima morning: a heavy damp mist was blowing through the dirty streets, and I was eager to get above it into the mountains where the sun would be shining.

I was also vaguely aware that every day spent at sea level was a day knocked off the valuable high-altitude acclimatisation I'd gained during the trip to Huaraz. But having scoffed at Tim's Acute Mountain Sickness book, I hadn't registered that just a couple of days spent back down at sea level is enough to wipe out all the acclimatisation gained on previous high-altitude ventures. If I had known this crucial fact, the day's events might have unfolded very differently.

As expected, the coastal mist lifted by the time we reached Chosica, about a thousand metres above sea level and a strange sort of inland beach resort town for sun-seeking Lima residents. The Rimac River itself flowed alongside for a while, filthy and polluted here in its lower reaches. The landscape was still desert, the mountain slopes utterly bare and desolate, without a single blade of grass to break up the monotony of brown sun-bleached rock. But all along the valley floor agriculture was thriving, as irrigation channels brought vital water to large fields of sugar cane, chilli peppers and maize. Maycoln turned the radio up loud, and we bombed along, roaring past the labouring juggernauts grinding slowly up the carriageway ahead, belching out clouds of foul diesel smoke.

As we gained height, slowly – almost imperceptibly – the mountainsides were changing. First cacti appeared, standing singly like silent sentinels on the lower slopes. Then the first bits of dry scrub came into view, to be gradually replaced with tough, tussocky clumps of grass. Little by little, we were climbing out of the desert.

As the steepening mountain walls closed in around us, Lima's radio stations faded one by one into a crackle of interference. Without taking his eyes off the road, Maycoln banged a salsa tape into the player, and the loud music continued.

The change in altitude was already detectable. When we stopped at a ramshackle shop by the road to buy chocolate and biscuits for the day ahead, my legs felt wobbly and my head strangely light. Still, I knew this was normal. Although severe Acute Mountain Sickness can lead to death by pulmonary or cerebral oedema (fluid filling in the lungs or brain); in the early stages headaches and even nausea are normal – and though unpleasant, of no great concern. Peru's central highway leads straight up from sea level to an altitude of 4800 metres in a space of about six hours – and even hardened truck drivers occasionally have to lean out of the windows to throw up. But as long as major exercise is avoided, and time is taken to allow the body to adjust to the lack of oxygen before making any further altitude gains, severe – and life-threatening – illness can usually be avoided.

It was these last two rules that I was just about to break.

We turned off before the main pass, up a precipitous side valley towards the Yuracmayo reservoir, just as Joel Campos had described. This was typical highland country: cows and alpacas grazed by the road, tended by *campesino* women in brightly-coloured skirts and black trilbies. The road was bumpy and it took an hour to reach the reservoir and pass through the grim-looking adobe village which clustered above the dam wall. The lake was several kilo-metres long, and on each side the mountain tops had that telltale brown and dusty look suggesting they had only recently lost their snow and ice.

Maycoln took the jeep as far as it would go along the grasslands and the riverbeds above the lake but, only a couple of kilometres further up, the way was blocked by a steep rocky slope and it was time for me to get out and start climbing on foot. It was already midday – with time pressing and another seven or so kilometres to climb I estimated that I needed to be at the Sullcon glacier by 3 p.m. at the latest if I was to have any chance of getting down safely before dark. According to the map, I'd already reached 4400 metres in altitude. Only another 600 or so to go.

I began to climb briskly, panting in the thin air. Ahead were white snow-covered peaks, but I wasn't sure exactly from the map which valley I was meant to be heading into. My mind felt strangely sluggish – as if I was half-asleep or even drunk – and my fingers had difficulty controlling the pen as I made notes. In fact, my fingers, face and feet were all tingling with pins and needles, and were even beginning to stiffen up slightly. I knew it couldn't be the cold, since I was well-dressed and the temperature was some way above freezing. A warning bell began to ring, but with only a short distance left to go, the intriguing prospect of estimating glacial retreat on Lima's most important glacier, and only one day before I was booked to leave the country, I decided to press on.

By climbing standards, the going was fairly easy. The valley floor was almost flat for the first few kilometres and, although nobody was around, a couple of thatched shepherds' huts stood at the corner of one field. A group of horses, surprised by my sudden appearance, cantered off into the distance. Higher up I had to cross a couple of steep-sided gullies, and at one point clamber along some rather unstable scree, but mainly the slope was grassy and un-challenging – although with no path to follow it was difficult to be sure of the best route. I seemed to be making good progress, and hit the first mounds of orange-and-white moraine at around 1.30.

Even in my half-conscious state, I could tell that this was clear evidence of recent glacial activity: debris left by a large glacier in its retreat uphill. Although the slopes had green plants growing on them, the valley bottom was desolate, just mounds of gravel and small rocks. I plodded on for at least another kilometre upwards through the desert-like moonscape, at last catching my first sight of the Sullcon glacier around the corner of a protruding ridge. It was still a good deal higher up, and sat like a dirty grey-white pancake at the top of the valley. High clouds had rolled in, and thick soft flakes of snow – just a few at first – were beginning to swirl around me as I reached the front of the ice an hour later.

Fully aware that something was wrong, I sat down on a ridge of gravel in front of the glacier edge, and tried to get some breath into my body. Although I was unable to hold the map properly at the time, and so had more or less given up trying to confirm exactly where I was, I had already reached an impressive 5000 metres in altitude.

The location was well worth pausing over for its own

sake: in a unique quirk of geography the Sullcon glacier actually straddles the continental divide – the very spine of South America itself – water from two pieces of ice right next to each other in the glacier could eventually end up over three thousand kilometres away in a different ocean (westwards via the Rimac to the Pacific, and eastwards via the Amazon to the Atlantic). Seated on my little ridge of debris, I allowed myself the brief pleasure of putting one foot on either side of South America whilst I ate a piece of chocolate.

Looking back downhill, I estimated that – judging from the lack of plant growth on the lower moraines – the glacier had retreated by about a kilometre in the last few decades. The remaining part of the glacier was pretty flat, and although ringed by higher snow-covered peaks, the areas where new ice could accumulate would already be much smaller.

Unlike the Cordillera Blanca, where many peaks pass the 6000-metre mark, the Cordillera Central simply isn't high enough to keep its glaciers alive for much longer. The Yuracmayo basin, at the head of which I was sitting, has lost a quarter of its glacial area in the last forty years alone.[38] The Cordillera Central as a whole lost a third of its ice cover between 1970 and 1997. In volume terms, that's 811 million cubic metres of water – nearly three times the volume of Windermere, Britain's largest lake – no longer held in ice form in the 'natural reservoirs' above Peru's capital city.[39]

*   *   *

Having come all this way, it seemed a shame not to climb up onto the ice itself and so, with a lot of effort, I hauled myself up the grey rock-strewn glacier tongue and over onto its surface. The going seemed safe enough: without a covering of snow to disguise the crevasses I could simply walk around – or even jump over – the yawning cracks which had opened up as the glacier ground its way slowly downhill. Lots of grit and occasional loose boulders had melted out onto the surface, and little pinnacles of ice crunched and tinkled as I walked on them. Rivulets cascaded down in all directions, eventually joining bigger streams that I could hear rumbling at the bottom of the deepest blue crevasses.

Reaching a high point of about 5100 metres, I stopped to take a couple of photographs (fumbling with the shutter, as my claw-like hands still refused to function properly), hurrying to capture the scene before it was blotted out by the increasingly heavy snowfall. Then, realising belatedly that I'd already passed my self-imposed 3 p.m. turn-around time, I began to head down.

It was already too late. Leaving the glacier and beginning to scramble back down the orange moraines, I had only stopped for a second to adjust my rucksack straps when I felt myself beginning to lose consciousness. My legs buckled and I hit the ground heavily. The pain of landing was accompanied by the sudden panic-stricken realisation about the process that was now underway. I knew that if I fell unconscious for any length of time I would die – either from cold during the night or from Acute Mountain Sickness itself. It really was as simple as that.

As I lay there gasping, I couldn't quite believe I had been stupid enough to let it come to this: after all the dangerous glacier climbing I'd done with Tim, here I was lying sprawled uselessly on the ground, virtually paralysed and alone, 5000 metres up in the Cordillera Central.

I had broken every rule in the book. No one knew where I was; I had no way of summoning help; and my paralysis symptoms corresponded exactly with life-threatening high-altitude cerebral oedema. I tried to curse myself, but even my mouth wouldn't move. My hat, I noted feebly, had fallen off and rolled down to the bottom of the mound of moraine. As I lay there, the snowflakes settled and thawed gently on my face.

The remaining functional part of my brain then moved into survival mode. I really, really didn't want to die. Lying down meant that I probably would die, because I knew my only chance of survival was to descend as rapidly as possible. So I decided to get up.

First, I managed to kick my legs around so that I was lying downhill – hoping this would help the blood flow to my head. Adrenalin was coursing through my veins, partly because of the panic and partly as an automatic bodily response to the threat of fainting. Then, using my stiffened arms as props, I rolled over and somehow struggled to my knees. After this, with another huge effort, I clambered onto my feet and began to stagger drunkenly downhill, rucksack hanging jauntily off one shoulder. Gasping with exhaustion and the sheer difficulty of breathing, I shuffled down the piles of loose rock and gravel.

Unfortunately, in my hurry to descend, I quickly lost the route and found myself on the wrong side of the valley, trapped by the stream entering a sudden gorge. This meant I would have to ascend again – something I felt was going to be impossible with my depleted energy reserves. I wandered around in circles for a bit, dazed and utterly directionless. The fainting feeling came back again, and simply as a reason not to stop I began to stumble back uphill. Fortunately, this was the right decision – emerging high on the opposite side of the valley from where I'd gone up, I picked up an animal track and eventually came back down to the grassy valley floor. As I crossed a slope of enormous boulders, several viscachas – large, long-tailed animals rather like big rabbits – came out of their burrows to watch the strange creature lurch by.

My driver Maycoln was stretched out on the back seat when I arrived back at the car. He was in bad shape too, and had recently vomited – but he quickly assured me that he was still able to drive. Slurring my words, it was difficult to explain what had happened, but he quickly grasped the urgency of the situation and started the car to begin our descent.

Acute Mountain Sickness is a funny beast, and miraculously, it seemed, every single one of my symptoms had evaporated by the time we were halfway back down to Lima. In fact, we were both suddenly ravenous, and stopped at a roadside restaurant for a large helping of rice and chicken. I felt unbelievably, preposterously lucky: and for days afterwards kept breaking out into sudden smiles of gratitude. It had been a very close shave indeed.

Back at home, my parents were not impressed at all.

'Oh, Mark,' wailed my mother, 'how could you be so stupid? If anything had happened to you . . .'

My father was more sanguine. 'I'm just glad you came back alive,' he said simply. I knew that he'd taken risks himself when he was mapping in Jacabamba, and that his disapproval was tempered by his own memories of clambering unassisted across the glacier to Nunatak Ridge.

That reminded me. I still had something to show him.

First I made sure we were all equipped with strong gin and tonics. Then I loaded my slides of Jacabamba into the projector and closed the living-room curtains on the darkening Welsh countryside outside.

First, I showed him the picture he'd taken in 1980 of the fan-shaped glacier above the lake. His eyes sparkled at the recollection. Then I pressed the forward button and up came Tim's photo, snapped only two weeks previously – of the same spot, with just bare rock and the glacier gone.

'Good God!' My father leant forward. 'I can't believe it. It's so sad. That was the whole *character* of the place. It was such a beautiful glacier, I've never forgotten it.'

He paused, trying to grasp the scene. 'Go back again, I want to have another look.'

His original slide came up, slightly bleached over time, but with the same glacier intact and healthy. I pressed the button again and there was Tim's photo, with nothing there but rock.

My father still couldn't quite believe it. The reality of the scene etched in his memory had passed away, and the world was the poorer for it. Today, the only remaining trace of the fan-shaped glacier above Jacabamba lake is a single fading impression on a strip of celluloid, and the similarly-fading recollections of a small group of geologists.

My father leant back again in his chair, and the sparkle in his eyes was gone. 'It's so sad,' he said again.

7

Feeling the Heat

It was late afternoon on Wednesday, the 22nd of November. Behind closed doors, in a large back room at the International Congress Centre in The Hague, Holland, a group of delegates and technicians were discussing the Kyoto Protocol. For nine days they had been closeted away from the outside world, speaking the jargon-encrusted language of 'compliance provisions', 'flexible mechanisms' and 'joint implementation'. But things were about to get a lot more interesting. Some uninvited guests had just entered the building.

'The delegates just sat there completely stunned when we burst in,' one of the protesters, George Marshall, told me. 'I think they were all brain-dead anyway, they'd been there for days.' The twenty activists didn't just hold up banners or hand out leaflets – they stormed the stage, sitting down and linking arms around the Chairman's top table and taking turns to berate the gathered technocrats for turning the talks into a 'carbon casino'.

Security guards rushed in and tried to drag the activists

away. Chaos ensued. At one point a table came crashing down.

Outside the closed room, another disturbance had begun: in the main foyer of the conference hall, a different group of activists had climbed onto an overhead beam, dropping banners and shouting, 'Climate talks, money talks. All talk and no action.' Bemused delegates peered up at the ceiling, where the protesters were staying firmly out of the reach of the lunging security guards. Several were swinging their legs and looking exaggeratedly bored. 'Blah blah blah blah,' they kept shouting. Due to some bureaucratic bungle, internal cameras were broadcasting the loud protests from every television monitor in the building.

Jan Pronk, the mild-mannered Dutch politician in charge of the whole conference, called a meeting of environmental groups and asked for someone to explain what it was all about. Melanie Jarman, a British activist, stood up.

'The chilling reality is that the climate summit in The Hague is close to an agreement that will not only enable Northern [rich world] governments and their corporations to escape their promised CO_2 reductions, but will allow them to significantly increase their emissions,' she said, reading from a prepared statement by the international grassroots coalition behind the protests, Rising Tide. 'This calamitous scenario will unfold if the wide range of fraudulent "solutions" promoted by some Northern governments and the unified corporate climate lobby are written into the Kyoto rulebook.'

And then, after a few more sentences, having made their point, the activists gathered up their belongings and simply walked out.

The Hague climate conference had already been going on for more than a week, and what really concerned the protesters and many of the other environmental groups present, was the apparent lack of urgency in the discussions about global warming. The international process had already dragged on for years, and everyone knew that the measures being talked about behind closed doors in the Congress Centre would only have a negligible impact on the real-world outcome. They might as well have been discussing a different planet for all the good it would do.

Back on Earth, meanwhile, evidence that the crisis is much more serious has continued to mount. Carbon dioxide is accumulating in the atmosphere twice as fast as natural processes can remove it. Carbon concentrations in the atmosphere are now higher than they have been for the last 420,000 years, and probably the last 20 million years.[1] The atmosphere we breathe today is chemically different from that breathed by our ancestors throughout the entire evolutionary history of the human race.

It's hardly surprising that as a result the Earth's climate has begun to tilt out of balance. The 1990s were the warmest decade since records began, and are likely to have been the warmest for the last thousand years and probably longer.[2] The top five warmest years ever recorded almost run in sequence: 1998 was the warmest, closely followed by 2002, 2001, 1997 and 1995.[3] Our planet is heating up rapidly, and the rate of warming is accelerating.

Computer model projections of climate change over the next century indicate that the speed and scale of warming could quickly become catastrophic. The Intergovernmental Panel on Climate Change has outlined a range of scenarios, which predict a temperature rise of between 1.4 and 5.8°C.[4] Whilst the lower end of the spectrum is still more than double what we have already experienced during the twentieth century, the higher end of nearly six degrees would take the Earth into uncharted and very dangerous territory indeed.

Many of the resulting impacts are predictable, such as the eventual melting of the ice caps, accompanied by an accelerating rate of sea level rise. According to the IPCC, hurricanes will probably get stronger, and the intensity of floods and droughts will increase too. Tropical diseases will spread towards the poles, and billions of people will begin to lose their water supplies. Ecosystems will unravel as plants and animals struggle to adapt to rising temperatures and to migrate fast enough to stay within their natural climatic zones. Agriculture will suffer, and food supplies will be endangered.[5]

The impacts will be disproportionately felt by the poor and vulnerable in semi-arid and tropical countries, many of whom are already living on the barest margins of survival. Conflicts over scarce resources become ever more likely, as do large movements of environmental refugees, when millions will be made homeless by extreme weather and seawater flooding of low-lying areas.

There is some uncertainty about the extent and speed of future warming and the consequent severity of these impacts, as the range of IPCC scenarios indicates. Although no one knows for sure just how sensitive the atmosphere is to increased greenhouse gas concentrations, most of this uncertainty focuses not on the science itself but on the behaviour of human civilisation. Will we remain dependent on fossil fuels for the foreseeable future, and continue to pump out billions of tonnes of carbon dioxide on an annual basis? Or will we set ourselves on a transition to a clean energy economy?

Added to this are dangers so difficult to quantify that the IPCC wasn't able to include them in its scenarios. For example, huge quantities of methane – more than double the world's entire fossil fuel reserves – are locked up in undersea deposits around the globe, kept stable by low ocean temperatures and the pressure of the water above. As sea temperatures creep up, large quantities of this methane (a potent greenhouse gas, twenty times more powerful than carbon dioxide) could suddenly enter the atmosphere and trigger a runaway global warming that we would be powerless to turn around. This kind of 'methane burp' seems likely to have triggered past episodes of rapid global warming during the Earth's geological history.[6]

Research by the UK's Hadley Centre has also raised another possibility: that drought-stricken Amazonian forests will begin to die back after about 2050, releasing all their locked-up carbon into the atmosphere. This process, as well as spelling doom for these richly biodiverse tropical ecosystems, would raise eventual global warming by an additional 1.5°C.[7]

And the bad news just keeps getting worse: every time a new scientific report comes out, it seems to revise the estimates of the Earth's climate sensitivity upwards. In May 2003 the Hadley Centre released the results of its most comprehensive computer model, which integrated just about every aspect of the global climate into a complete 'Earth-systems' approach. The central estimate of the planet's response to a 'business as usual' emissions scenario was raised from 4°C to 5.5°C, perilously close to the IPCC's previous 'upper limit', which may now have to be raised still further.[8]

Hot on the heels of this, came another piece of news, this time from a meeting of top climate scientists in Berlin. They concluded that previous models had underestimated the cooling effect of smoke and other atmospheric particles, reducing the past extent of global warming by up to three-quarters. So rather than the 0.6°C of warming experienced over the past century, we would have had 1.8°C if it hadn't been for the smoky results of forest destruction and coal-burning power stations. With smoke pollution now in decline due to concerns over its health effects and acid rain, the true extent of warming will hit hard over the next century – possibly raising global temperatures by a staggering 7 to 10°C.[9]

But however uncertain the eventual extent of warming,

it's crucial to remember that the *least* likely option of all is no change. The world will not continue to enjoy a fairly stable climate, as it has through centuries of human history. Some warming, probably double what we've experienced so far, is inevitable.

If this temperature rise is not to continue accelerating towards these catastrophic levels, urgent action is required – the kind of action which would see all the world's nations sitting down at a table and agreeing massive cuts in greenhouse gas emissions.

But anyone who hoped for this outcome from the Hague meeting was about to be sorely disappointed.

At issue during the Hague negotiations were not big cuts in greenhouse gas emissions but the arcane rules for implementing the Kyoto Protocol, agreed back in 1997. The participants at Kyoto had agreed that industrialised country greenhouse gas emissions would be cut by a total of about 5% below 1990 levels by 2012. Although this was obviously a far cry from the long-term 80% cuts that most environmental groups wanted, it was at least a start.

However, by November 2000, Kyoto's emissions reductions, tiny as they were, had still not come into force. Every year negotiators from all the world's nations would reassemble in an attempt to finalise the small print of the Protocol's mechanisms, but every year they would be faced off by a coalition of countries called the Umbrella Group – led by the United States, but also including Australia, Canada and Japan.

Egged on by representatives of some of the world's most powerful fossil fuel corporations, the Umbrella Group could also count on vocal support from the oil-producing countries of Kuwait and Saudi Arabia. As a result, there was a big question looming over the Hague conference: would the Kyoto Protocol survive these co-ordinated attacks, or would it unravel and eventually be abandoned?

Kyoto, battered and inadequate as it was, had been the product of a long and tortuous negotiating process. It was only agreed after a desperate battle by environmental groups and progressive countries against the alliance of powerful corporations, OPEC oil-producing nations and the United States – as well as a great deal of political horse-trading and a last-minute personal intervention by the then US vice-president Al Gore.[10]

Kyoto was the outcome of an even longer process, dating back to the Rio Earth Summit in 1992, when the United Nations Framework Convention on Climate Change (UNFCCC) was signed by world leaders – including then US president George Bush Snr.

The Convention's widely-quoted Article 2 is still the benchmark for international action to combat climate change. It committed the international community to the 'stabilisation of greenhouse gas concentrations in the atmosphere at a level that would prevent dangerous anthropogenic interference with the climate system'. The UNFCCC looked like a big success for the environment, especially as rich countries were expected to freeze green-house gas emissions at 1990 levels by the year 2000.

But the Convention had no teeth. The stabilisation commitment would only be a voluntary one, thanks yet again to intervention by the United States – Bush even threatened to boycott Rio altogether if the Convention included any mandatory targets. And then, as tends to be the way with voluntary agreements, it was immediately forgotten about: everyone simply went home and got on with business as usual.

The Kyoto Protocol was meant to be the first of the Convention's teeth, forcing countries to take joint action where the voluntary approach had so clearly failed. But a number of loopholes had been carefully winkled open by the Umbrella Group over the years, and by the time of The Hague these were collectively posing such a threat that Kyoto's mandated 5% cut in emissions was in danger of becoming symbolic rather than real.

One of the biggest was the issue of 'carbon sinks': forests and tree plantations which are assumed to be absorbing carbon dioxide from the atmosphere and can therefore be counted against a country's emission reduction target. In the logic of sinks supporters, simply planting (or not destroying) forests was the same as cutting back on oil or coal consumption.

If it sounds too easy to be true, that's because it is. For one, it's almost impossible to tell how much carbon dioxide a tree is actually absorbing – forests both absorb carbon in their leaves and release it from soils at widely varying rates depending on many different factors. Australia, despite being one of the world's biggest emitters of greenhouse gases, was insisting that 'revegetation' should be counted towards its target – without being very specific as to which vegetation it meant, whilst continuing to deforest vast areas of old-growth woodland. Members of Friends of the Earth illustrated this by taping signs declaring 'this is not a sink' on potted plants all around the conference centre. They also left a real (kitchen) sink nearby, next to a message which reassured confused delegates that this *was* indeed a sink.

Carbon which has been buried underground for millions of years as coal or oil cannot be considered the equivalent of carbon locked up in trees for only a matter of decades. If 'sinks' became widely accepted, humanity could go on digging up fossil carbon and adding it to the living biosphere, where much of it would eventually end up in the atmosphere, just as before.

Another controversial loophole was whether or not a country which had cut its greenhouse gases more than required (and was therefore under budget) should be able to sell the difference to a country in danger of overshooting the target. The biggest potential seller was Russia, whose economy (and therefore emissions) had collapsed since the 1990 baseline, and which would therefore have enormous numbers of emissions 'credits' to sell on the open market. If all Russian 'hot air' were sold, other industrialised countries like the US and Japan might not have to make any carbon cuts at all.

Emissions trading had turned into one of the biggest bugbears for the more radical environmental activists, who were sceptical of market-friendly solutions in general and particularly of the private sector corporations that stood to gain from them. Not only could trading mean that real reductions simply evaporated into hot air, it could also lead to the creation of huge new markets, with 'carbon credits' being traded and speculated on just like any other sellable commodity.

As the protesters had told Jan Pronk during their earlier invasion of the conference:

> People are suffering from the operations of the oil multi-nationals and the impacts of climate change, right now. And what they are seeing here [in The Hague] are those same companies which destroyed their lands and cultures con-spiring to profit from the very problem they have caused. Yet the plight of these people isn't even on the agenda. Instead we have a carbon casino masquerading as a solution to a global crisis. We need drastic cuts in carbon emissions – sixty per cent worldwide, eighty to ninety per cent in devel-oped countries – to halt climate change. But what we see at [this conference] is a huddle of gangsters, plotting the most profitable scams to dodge even the woefully inadequate Kyoto targets.

That evening, at the usual press briefing by the United States delegation, the activists struck again.

The US briefing was a must-see: every evening the dele-gation leader Frank Loy would laconically outline all the ways in which the United States was trying to secure the best possible agreement to tackle global warming.

In reality, he meant exactly the opposite. For the past week the US had been doing its utmost to undermine even the very unambitious climate change agreement which was on the table. Loy was clearly unmoved by the fact that the US is the world's biggest polluter – America, with only a twentieth of the world's population, is responsible for a quarter of all fossil fuel emissions.

I was sitting on the front row, four seats away from a smartly-dressed young woman, when the US delegation head Frank Loy walked in with his team and took his place. The young woman lifted up her bag and walked quickly to the front table where Loy was arranging his briefing notes. She then put her hand inside the bag. Loy, assuming that she was just another journalist placing a tape recorder on the desk to record his comments, ignored her.

Instead she pulled out something white, and in a flash had pushed it into his face – it was a very messy cream pie. At the same instant another woman further back in the room stood up and shouted: 'This is all a farce!'

Pandemonium broke out. Security guards rushed over and seized the pie-woman. Photographers leapt over chairs to get close-ups of Loy as he bemusedly wiped cream off his glasses and face. The other US delegation members (as soon as they regained their composure) began rushing around looking for napkins to assist in their leader's clean-up process.

To his credit, Loy didn't lose his cool. Instead, he grinned, leaned forward to the microphone, and began to speak – bits of cream dripping from his face as he did so. 'On the eve of Thanksgiving,' he drawled, 'pumpkin pie would have been a more traditional choice, but what I really want is a strong agreement to fight global warming. I'm headed back to the negotiating table right now with that aim.'

Then he walked out and the press conference was over. Several reporters clapped, impressed by Loy's performance and angry that the pie-woman had robbed them of their chance to ask their daily quota of questions. But outside the conference room the two female activists had also scored their own victory, managing to shake free of their guards and escape into the night.

'This is all a farce!' one had shouted. And it was difficult to disagree. On the day before the conference was due to close, Jan Pronk had produced a 'compromise document' aiming to bridge the gap between the loophole-supporting Umbrella Group, and the European Union, the G77 group of developing countries and China – which supported tougher action.

A quick analysis showed that the proposed deal was hopelessly biased towards the demands of the Umbrella Group, much as the activists had predicted. Pronk had given in to the US demand that large amounts of carbon dioxide supposedly sequestered in America's forests and farmland count towards its Kyoto targets. He had also taken out the cap which would have limited emissions trading – the giant market in 'hot air' trading would go ahead unimpeded.

The proposed agreement neglected to rule out nuclear power as a way for countries to help each other meet their targets (and trade the difference), whilst allowing rich countries to establish massive forestry plantations (with no environmental safeguards, this potentially leading to the destruction of natural forests) in poor ones as a way of getting out of their domestic commitments to reduce emissions.

The response from mainstream environmental groups was understandably angry. Friends of the Earth's Roda Verheyen told journalists:

> The Kyoto Treaty was the bare minimum required to tackle man-made climate change. The latest attempt to get a deal in The Hague represents a further watering down of this treaty. The US and its supporters have been given a huge free gift in the form of carbon credits towards their Kyoto targets and the right to trade vast amounts of hot air. This means they can avoid action at home to cut emissions from fossil fuels. Man-made climate change will not even be stopped by this deal, let alone reversed.

The World Wildlife Fund's Jennifer Morgan was also furious. 'Pronk's paper places economic convenience over the protection of the planet. This text, as it stands, will allow emissions to increase rather than ensuring that they are reduced.' Greenpeace, for its part, put its experts onto crunching the numbers – and they had concluded that if Pronk's deal was agreed as it stood, the completed Kyoto treaty would result in emissions *increases* of between 6 and 9% rather than the intended reductions of 5%. 'The European Union must fight,' Greenpeace declared.

As the final hours ticked away that evening, the French Environment Minister Dominique Voynet (herself a Green Party MP) told a packed press conference that the Europeans would stand firm. 'We must not lose sight of the objective,' she declared, reminding the assembled journalists that at Kyoto three years earlier it had taken lengthy debate to agree on the 5% emissions reduction, 'and we're not here to undo all that.' As a result, she went on, 'we want to guarantee that the commitments made in Kyoto are actually carried out. The [Pronk] document does not meet these requirements.'

The EU seemed to be agreeing with Greenpeace. 'This compromise is totally unbalanced from the point of view of environmental integrity,' Voynet concluded, adding:

> This would turn the agreement into one that would authorise an increase in emissions, and would open [loopholes] which would be very difficult to close in the long-term. We are not prepared to rubber-stamp a bureaucratic set up which would not make it possible to abide by the commitments.

The United States, cancelling its usual evening press conference, declined to comment. But in a terse written statement, its head negotiator Frank Loy also declared himself 'deeply disappointed with Minister Pronk's paper, which we consider unacceptably imbalanced'. Quite why, given that the US had appeared to be the big winner, he neglected to say.

To be fair to Jan Pronk, it was his role as conference president to suggest compromises which might bring the various parties together into a final agreement – whatever his personal views. And he'd made his own commitment to the Kyoto Protocol very clear, even speaking at a Friends of the Earth rally – where thousands of supporters built a symbolic dyke out of sandbags – outside the building earlier in the week. 'The negotiators are trying to build a dyke with words,' he had told the crowd, whilst holding up a single sandbag which he promised to take inside and keep on his main podium desk. 'We need to build a dyke with action. The message is – don't destroy the climate, and don't destroy the climate treaty!'

But those hopeful words had so far led to very little. And even in the press room, that final evening, feelings were running high. Announcing their presence with a loud bugle call, two Canadian journalists burnt their passports in protest against their country's role in supporting the US. Spokespeople from environmental groups stalked up and down, occasionally distributing hastily-photocopied press releases. 'No deal is better than a bad deal,' announced Greenpeace's Bill Hare, a climate negotiations veteran, to one group of reporters. Other observers, meanwhile, prepared to wait it out, settling down in quiet corners to catch a few hours' sleep. A few hardy journalists, anticipating the long night ahead, retired to the bar.

Some seasoned conference-watchers were still guardedly optimistic: 'They always stitch something together in the small hours of the morning,' one British journalist had reassured me. Many felt that the US would back down in the end, especially as this would be one of President Clinton's last acts before handing over power to the newly 'elected' George Bush. Surely Clinton's vice-president Al Gore, the man who was once so strongly committed to environmental issues that he penned a book entitled *Earth in the Balance*, would want to make his mark on history whilst he still had the chance?

But time was running out. And behind closed doors, the endgame had begun.

It was indeed a long night – and by the time I returned at eight in the morning, there was still no deal. Rumours flew around, and in the press room small groups of baggy-eyed hacks occasionally gathered to compare notes. At one point a journalist from the *Independent* claimed to have details of an agreement – even down to the small print on sinks and emissions trading. It seemed that the UK's strident deputy Prime Minister, John Prescott, was on the verge of hammering out a last-minute face-saving deal between the European Union and the United States.

But everyone had a different view: even the environmental groups had a trans-Atlantic rift, the European and international groups claiming that the Americans were still holding out, whilst the US-based green groups blamed the European Union for refusing to compromise. However, even late into the morning, no one knew for sure whether Prescott would succeed or not.

The answer came at 11.35 a.m. precisely. I was in the main foyer when John Prescott himself strode past, trailing a gaggle of excited journalists.

No deal.

'I'm gutted that we didn't make it, particularly for all the people who wanted change,' was all that he would say. 'But that's life.'

I caught up with him outside the conference centre, and asked breathlessly: 'Are you walking out?'

'I haven't walked out,' he replied tersely, nearing his hotel. 'It's the end of the conference.'

Everyone looked shell-shocked. Even the most pessimistic observers had expected some kind of compromise to be struck. Now there was nothing at all.

Outside the building Friends of the Earth activists began to angrily push their dyke into the pond. Their banner said it all: 'You've sunk the Protocol'. In the press room a final Friends of the Earth press release was circulating. 'Hot air!' ran the headline. 'Two weeks of talk: US ensures no result.' In a hastily-called media briefing, Roger Higman was in no doubt about who was responsible. 'The US has wrecked the spirit of the Kyoto Protocol and undermined action to tackle climate change. Make no mistake, the US is to blame.'

His colleague Tony Juniper added eloquently: 'The failure of these talks is a disaster for us all. No words can truly express our anger at what has happened here, or our sadness for the victims of the climate change that is to come. The world will pay the price in tears.'

Greenpeace was just as unforgiving: 'This meeting will be remembered as the moment when governments abandoned the promise of global co-operation to protect planet Earth,' said Bill Hare.

But Frank Loy saw things somewhat differently. 'The United States put forward a number of constructive proposals in hopes of arriving at a more positive outcome,' he said in a printed statement (there was no final press conference). 'The elements of an agreement were sitting on the table in plain sight. It included some of the toughest issues – sinks, compliance, and ensuring strong domestic action. We were ready to sign on to that agreement – others could not.'

In Loy's view, the European Union had blocked the way forward, by sticking too tenaciously to its position.

I am troubled that some of our partners in these negotiations held fast to positions that not only strayed from the Kyoto bargain, but ignored some of the fundamental realities we face. They ignored environmental and economic realities, insisting on provisions that would shackle the very tools that offer us the best hope of achieving our ambitious target at an affordable cost.

He listed carbon sinks and emissions trading as two 'legitimate means' of meeting America's Kyoto targets which 'our negotiating partners' (presumably the EU) had refused to accept. 'The United States will continue to be a leader in this fight [against global warming],' he promised, but it 'is not in the business of signing up to agreements it knows it cannot fulfill. We don't make promises we can't keep.'

Bill Clinton had missed his chance.

Jan Pronk had the unenviable task of winding up the conference. He looked tired and defeated. 'Personally I am very disappointed,' he admitted to the remaining assembled delegates, many having already left to catch their flights home. 'We should be aware of the fact that we have been watched by the outside world. There have been expressed high expectations, and I must confess today that we have not lived up to the expectations. That is a fact.' He sighed and, rather surprisingly, read out a poem.

> All of our dreams, our mass-dream-spinners,
> Palaces, hovels, every lair –
> Seem, when the Flood comes, to be
> Just castles in the sand or air.

A bemused silence followed. Then the delegate from Antigua stood up. 'These negotiations need not die,' he said. 'I propose that we suspend the negotiations and resume in Bonn.' It seemed like a sensible suggestion – Bonn was the home of the Climate Change Convention's Secretariat. Argentina's Ambassador Estrada, a former conference chairman, agreed. 'I and other members are in favour of suspending and resuming in Bonn. The UNFCCC is a process – it doesn't end. Each meeting bears its fruits.'

Hungary, Nigeria, Norway and many other countries lined up to support the proposal. This was not the end, they agreed. The conference, and by extension the Kyoto Protocol, would get a second chance – in reality a last chance – the following year at Bonn. That meeting would be a very different one, and the stakes would be higher still.

BONN, JULY 2001

It was difficult to believe that things could get much worse. US obstructionism had almost sunk the Kyoto Protocol – even in the last days of an administration whose vice-president Al Gore was once a self-declared green. Once George W. Bush arrived in office, things degenerated even more rapidly. Clinton had at least negotiated, but Bush – with the arrogance that quickly came to characterise his presidency – went much further, declaring on 28 March 2001 that he was taking the United States out of the Kyoto Protocol altogether.

The new president made no bones about the fact that he hoped the American move would destroy the Kyoto Protocol, and by implication international efforts to control emissions of greenhouse gases, altogether. 'The President has been unequivocal,' the White House press corps were informed. 'He does not support the Kyoto treaty. It is not in the United States' economic best interest.' At a press conference the following day, Bush added: 'I will not accept anything that will harm our economy and hurt our American workers.' Vice-President Dick Cheney made his own position abundantly clear a few days later. 'The Kyoto pact is dead,' he told ABC television.

Cheney himself had already taken the leading role in forging the new administration's energy policy, chairing a high-level group which reported in May 2001. Its conclusions were climate-unfriendly to say the least. At least 1300 new power plants, mostly fuelled by coal and gas, would be built over the next twenty years. There would be no regulation of carbon dioxide emissions from these new plants, nor any of carbon emissions from the existing 1082 coal-fired plants, which produce the majority of America's electricity. Although coal is the most carbon-intensive fuel of all (i.e. per unit of energy produced, coal releases the most carbon dioxide, followed by oil and then gas), there would be a massive $2 billion subsidy for new coal-based technologies.

The plan did nothing to increase the fuel economy of America's 200-million-strong fleet of cars, SUVs and trucks, nor to reverse the weaker efficiency standards Bush had already announced for air-conditioning systems – a measure which in itself would force the construction of at least forty new power plants by 2020, and generate 180 million more tonnes of carbon dioxide over the next thirty years.[11] Instead, it proposed to increase domestic oil drilling, primarily by opening up the Arctic National Wildlife Refuge and leasing new drilling areas in other public lands such as Wyoming's Red Desert and the Rocky Mountains, as well as off the coasts of Florida and Alaska.[12]

Although Cheney has tried to keep secret exactly who had access to his Energy Task Force, one environmental group has discovered that industry representatives enjoyed close access to the policy-making group, having twenty-five times as many direct contacts with it as did non-industry groups.[13]

This clear bias fuelled allegations that it was 'payback time' for the corporations which had poured money into Bush's presidential campaign. The donations had indeed been significant, with the oil and gas industry putting $1,900,000 behind the Republican 2000 presidential ticket. Their gift was nearly matched by $1,270,000 from the automotive industry. (Bush's opponent Al Gore, on the other hand, received only $340,000 from the energy sector.)

The now-disgraced Enron was one of the top contributors, whilst Exxon – the oil company which has most consistently opposed action to tackle climate change – also ranked highly, giving over $1 million in political donations during the 2000 election cycle, 90% of that to the Republicans.[14]

Many people have observed that the Bush administration bears more resemblance to an oil mafia than an elected government. Not only are there direct corporate interests at stake, but the entire mentality of the Bush administration seems to be based on the perception that what is good for oil corporations is good for America – and, by extension, the world.

Bush himself, of course, is a former Texas oilman – albeit a rather unsuccessful one, having been rescued from near bankruptcy by his father's connections.

The President's most favoured lieutenants are also

closely connected with the fossil fuels industry. Before joining the government, Dick Cheney was Chief Executive of Halliburton, the world's largest oilfield services company. It is thought that Cheney amassed some $50–60 million there, riches which stand out even in a cabinet dominated by millionaires.[15]

Bush's Commerce Secretary Don Evans (whose brief includes oversight of the National Oceanic and Atmospheric Administration, home of government-sponsored climate science) was recently Chairman and CEO of Tom Brown Inc, a $1.2 billion Denver-based oil and gas company. Energy Secretary Spencer Abraham has close ties to the automotive industry, and during his unsuccessful run for the Senate in 2000 was the number one recipient of car company handouts.

The President's Chief of Staff, Andrew Card, was formerly chief political lobbyist for General Motors, and before that headed the main car industry lobbying group, the now-defunct American Automobile Manufacturers Association. Much of Card's previous work had involved fighting on behalf of the biggest US automakers against higher environmental standards.[16] National Security Advisor Condoleezza Rice was formerly on the board of directors of the oil company Chevron, and was so helpful to the company that it named a 130,000-ton oil tanker after her.

Bush's repudiation of Kyoto caused dramatic international repercussions. Reminding the US that it is the world's most polluting nation, Britain's deputy Prime

Minister John Prescott insisted that America 'cannot pollute the world while free-riding on action by everyone else'.[17]

'Kyoto is still alive,' insisted Sweden's Environment Minister Kjell Larsson. 'No country has the right to declare Kyoto dead.'[18] France's Dominique Voynet went further, calling the US move 'completely provocative and irresponsible' and warning Bush not to continue the 'sabotage' if other countries persevered with Kyoto.[19]

Only Jan Pronk offered an olive branch. 'I think it is to a certain extent youngness of the administration and over-emphasis on domestic policies rather than on international policy-making,' he suggested generously. 'I would say give them some time to get their act together.'[20] Perhaps in this spirit an EU team flew to Washington, but returned – to the surprise of no one except themselves – empty-handed and ignored.

For a while it seemed as if the world might unite against America, but by the Bonn conference, a different kind of politics was in play. The European Union was now on the defensive, and the remaining members of the Umbrella Group such as Canada, Australia and Japan (now minus their American leader) as a result found their position much stronger. Knowing that the Europeans wanted desperately to save Kyoto after Bush's snub, the Umbrella Group was free to press for greater concessions in loophole areas like sinks, emissions trading, and compliance ('compliance' dealt with the penalties for countries which failed to meet their Kyoto targets).

A new period of brinkmanship was opening up. When I arrived in Bonn, the question was whether the EU would back down far enough to save Kyoto, and if it did, whether the weakened Protocol would be worth the paper it was written on.

Even though America had repudiated the Kyoto Protocol, it remained a party to the Climate Change Convention, and had sent a delegation accordingly. The shadowy presence of the world's most powerful nation still loomed over the entire proceedings. And rumours immediately began to fly about back-room pressure being leveraged by the US delegation, despite a promise by its leader Paula Dobriansky to the conference chairman Jan Pronk that the US would stay out of Kyoto-related discussions.

The sour-looking Dobriansky didn't help her case by making a transparently hypocritical speech at the first plenary. 'The United States takes the issue of climate change very seriously,' she began – without a hint of irony, to astonished silence from the other delegates. 'President Bush has pledged that this administration will play a leadership role in addressing this environmental challenge.' Dobriansky repeated the promise she had made to Pronk: 'While we do not believe the Kyoto Protocol is sound public policy for the United States, we do not intend to prevent others from going ahead with the treaty, so long as they do not harm legitimate US interests.'

The Eco conference freesheet, produced by a coalition of environmental groups, parodied Dobriansky's speech simply by editing it. 'I . . . look forward to . . . climate change. We are taking . . . more. We are moving actively to enhance . . . climate change. Over the next few days . . . the United States will . . . abdicate our responsibilities. Thank you.'[21]

But the next couple of days were no laughing matter. It became clear that the European Union was no longer holding firm in its position against sinks and seemed to be giving ground on emissions trading. At a 'technical briefing' that evening, the EU delegates admitted that they were simply trying to reduce the 'scale' of the sinks giveaways demanded by Japan, Australia and Canada. And on 'compliance', Japan in particular was holding out against the idea that countries should face binding penalties if they failed to meet their Kyoto commitments. Even the usually unruffled Jan Pronk seemed to find this hard to take. 'It seems that something about the whole compliance issue is a problem,' he said exasperatedly on the final day. 'But how can you have an international agreement that no one has to comply with?'

Again to try and break the deadlock, Pronk produced another compromise paper – but this time, unlike in The Hague, he seemed to have got it right. Although the plan did allow forestry 'sinks' to be counted against emissions, it also proposed a binding compliance regime and excluded nuclear power. This was enough for the environmental groups, who lined up in reluctant support. 'There is momentum building now,' said WWF's Jennifer Morgan. 'Japan, Canada and Australia have now won all the sinks ground they wanted. There is no reason for the US to block this package. Now is the time to strike the deal.'

But whether a deal would be struck remained up to governments, not campaigners. As the last hours ticked by on the final evening, there was little to do but join the crowds in the bar, where green campaigners and corporate representatives wove in and out of groups of journalists passing out press releases and bits of conference gossip. A short distance away, Don Pearlman – whose tireless work on behalf of the fossil fuel lobby had taken him to every climate conference for at least the last ten years – was briefing negotiators in an urgent whisper. I sidled over and took his photo, which seemed to annoy him.

In the corridor outside, the hubbub rose and fell – but the chairs around the crucial table remained stubbornly empty, as all the negotiations were now between small contact groups and behind closed doors. A little later Jan Pronk arrived in the main plenary hall to announce a new deadline – 4 a.m. There was an immediate rush amongst the crowded observers to try and find places to sleep: I managed to bag a leather couch at the back of the plenary room. Within minutes the whole area was transformed into a virtual dormitory, with reporters, lobbyists and exhausted delegates stretched out and snoring peacefully on chairs, sofas and even the carpet.

But there was no plenary at 4 a.m., and nor was there any news when I joined the crowd of people shaving and brushing their teeth in the conference toilets at nine. The latest rumour was that only the Japanese were now standing in the way of an agreement by objecting to a single sentence on compliance. But giving in was not an option, because it would make Kyoto essentially a voluntary agreement – and we all knew what that meant.

I was in the main foyer again when the hubbub suddenly changed. Within seconds, it seemed, everyone was smiling. They'd succeeded. The Bonn Agreement would happen. Next to me two of the key EU delegates were hugging each other.

'We did it!' they beamed.

Although details were sketchy at first, the scale of the EU's retreat soon became clear. Countries would be able to count far more 'sinks' activities – whether forest, cropland or grazing land management – than previously. Emissions trading was still supposed to be 'supplemental' to cuts at home, but there would be no formal limit. In terms of 'compliance', countries which overshot their Kyoto targets would be forced to make up the shortfall, plus 30%, in the next target period (after 2012), as well as being ruled ineligible to sell credits and having to draw up a 'compliance action plan'.

It was a mixed bag but much better than nothing, and it did seem to mean that the long-delayed Kyoto Protocol could at last be ratified by the countries who still supported it and come into effect. Emissions were still soaring, and now even Kyoto's tiny 5% reductions were beginning to look radical.

Just minutes after news of the deal came through, the European Union called a press conference. The EU ministers were flushed and smiling as they filed in. 'The rescue operation succeeded. We are very happy,' said Environment Commissioner Margot Wallström. 'Tired but happy. I think we can go home and look our children in the eye.'

Olivier Deleuze, the Belgian Energy Minister, admitted that the deal was imperfect, and that the EU would have preferred less sinks and a tighter compliance regime. They had spent the night haggling over just two phrases in the compliance section, he revealed. 'But I prefer an imperfect agreement that is living to a perfect agreement that doesn't exist.'

I asked him how he would rate the Bonn Agreement for environmental integrity on a scale of one to ten. 'It's a first step,' he answered. 'In comparison to the IPCC reductions it's one out of ten. But if you equate it with the need to make a first step, it's perfect.' Someone else asked what the message was to Bush. 'The message to President Bush is that he's welcome,' replied Deleuze, with an enigmatic smile.

The only thing left was the final plenary. It was already packed, but I managed to squeeze onto the balcony between a railing and a Japanese journalist. Down below all the delegates were shaking hands. I watched as the EU's Margot Wallström walked up to the Japanese Environment Minister Yoriko Kawaguchi and congratulated her, their beaming faces lit up by the glare from a circle of TV cameras.

Then Jan Pronk entered the room from a side door to take his place at the chairman's desk. As he did so, the German Environment Minister Jurgen Trittin stood up, raised his hands above his head and began to clap. Within seconds the whole room had joined in, giving Pronk a standing ovation. Pronk and Michael Zammit Cutajar from the Climate Change Convention Secretariat shook hands for a minute and then abandoned traditional conference reserve and hugged. All around the applause rose to new heights. Even the journalists and environmental campaigners were joining in. There was yelling and cheering, shaking the whole hall with jubilant noise.

'Good morning,' Pronk began, when at last quiet had descended and everyone had sat down. 'I now declare the sixteenth Plenary open.' He banged his gavel. 'Let me apologise for the delay,' he continued drily, to subdued laughter and a scattering of applause. 'We did have long and intense negotiations throughout the night, but I think it was most worth it. Negotiations on compliance concluded at ten this morning. I now propose that the conference approves this draft text. No objections?' He looked around the room. No one moved a muscle. 'If it is so decided,' Pronk said, and banged his gavel one final time.

Again applause broke out, the celebrations accompanied by relief that the humiliating failure at The Hague had at last been reversed – and the Kyoto Protocol saved from oblivion.

Then speakers from the floor took turns – first Ambassador Bagher Asadi from Iran, representing the G77 and China group of developing nations, stood up, then Olivier Deleuze on behalf of the EU, then Australia's Senator Hill for the Umbrella Group.

Then came the moment we'd all been waiting for. Paula Dobriansky, representing the United States of America, took the floor. 'The United States came to this global conference on climate change to work constructively to enhance international co-operation on an important environmental challenge,' she began, again without the tiniest hint of irony.

There were several meaningful coughs. Unsmilingly, she continued.

> Regarding the adoption of rules elaborating the Kyoto Protocol, although the United States does not intend to ratify that agreement, we have not sought to stop others from moving ahead, so long as legitimate US interests were protected. At the same time, the United States must emphasise that our not blocking consensus on the adoption of these Kyoto Protocol rules does not change our view that the Protocol is not sound policy. The decisions made today with respect to the Protocol, in addition, reinforce our conclusion that the treaty is not workable for the United States.

She went on to point out, line by line, elements of the Kyoto Protocol which the United States – despite no longer being a participant – still found unacceptable. There were muffled gasps of outrage all round. I looked at Pronk. He was giving nothing away, his face impassive.

But Dobriansky had overstepped the mark. 'The Bush administration takes the issue of climate change very seriously . . .' she began, before being interrupted. Boos, catcalls and yells of derision rang out from across the hall – the delegates' tables included. I let out my loudest yodel, whilst the Japanese journalist next to me did an impressive boo. Something had snapped, and all the world's injured pride and fury united everyone in the single objective of calling Bush's bluff.

Jan Pronk seemed to wait just a second too long before banging his gavel on the table to restore order. As he watched the scene unfold, I saw just the flicker of a smile play around the corners of his mouth.

THE WAY OUT

The fight to stop major climate destabilisation has only just begun, but the battle lines have already been drawn. On one side stands a wide-spectrum coalition of the environmental groups, the European Union, the non-oil-producing Third World countries and the small island states, together with an increasingly large and vocal groundswell of public opinion.

On the other stands the Bush administration and its allies like Australia and Saudi Arabia, major oil, coal and gas corporations, car companies and the ragtag bunch of 'climate sceptics' who struggle daily to undermine scientific knowledge and get their message that climate change isn't happening into the popular media. (The sceptics are completely absent from the scientific literature, where any debate about whether climate change is not happening is all but over.)

Ultimately, Bush and co. have to fail in this fight, for the sake of us, our children and the world's climate. And for that to happen, we *do* need to take sides. No one can be neutral in the struggle that lies ahead.

In the Appendix to this book, I give suggestions for other information sources, campaigning groups and websites that readers might find useful. But, first, I want to put forward some suggestions, which together add up to some sort of manifesto for how we might begin to find a way out of this crisis together.

1. Ratify and implement the Kyoto Protocol

At the time of writing, over two years after the Bonn Climate Conference, the Kyoto Protocol has still not come into force. The US is no longer alone in fighting it: Australia has now confirmed its position at America's side. But both Japan and Canada have signed up, and now only Russia still needs to come on board to trigger Kyoto's implementation.

It's true that Kyoto only mandates a tiny cut in emissions, at a time when massive cuts are needed. But its real value perhaps lies in the fact that it represents a long-term process for bringing all the world's nations together, a process which may in the end yield the kind of radical action which is clearly becoming necessary.

In the meantime, to begin the halting first steps towards protecting the Earth's climate, the Kyoto Protocol must be brought into force – and soon. We must urge the Russian government to ratify Kyoto, and other governments to implement it seriously once it comes into effect.

2. Sign up to 'contraction and convergence'

Only industrialised countries are active participants in Kyoto: developing countries have refused to take on their own cuts on the reasonable grounds that it could freeze their development and worsen global inequalities. But Third World countries account for an increasing share of global emissions: China is the second largest polluter after the US, and India is also in the top ten.

Clearly greenhouse gas emissions from the developing world will also need to be reduced soon if dangerous global warming is to be avoided. However, discussions on this have not even started, and attempts to begin negotiations at the Delhi Climate Conference in 2002 were rejected.

One crucial reason for this rejection is equity. Why should India and China – whose citizens on average emit respectively only a tenth and a quarter as much as the average British citizen – agree to limit their consumption now, when the industrialised world has got rich on the back of a century or more of carbon-based development? The issue is likely to cause deadlock for years into the future, unless someone can find a clever way around it.

Luckily, a workable solution is currently on the table, one which recognises that equal rights to the atmosphere are integral to efforts to protect the climate from major destabilisation. First developed by Aubrey Meyer of the Global Commons Institute in London, it has begun to receive tacit support from within the British government, adding to support from the European Parliament, the Africa Group of Nations and the governments of India and China. This solution has an elegant logic which cuts right through all the UN jargon and complexity which has blighted international climate policy so far. It's called 'contraction and convergence'.[22]

The way it works is simple. First, the world agrees an atmospheric greenhouse gas concentration target which will keep global warming within safe boundaries. This target then translates into a global emissions budget, which is parcelled out on an equal per capita basis across the world. Every Chinese, American, Bangladeshi and Greek would get the same entitlement, phased in over an agreed convergence period.

These entitlements should, Meyer insists, be tradeable – both to ease the transition and to generate much-needed revenue flows from rich to poor countries. (This will differ from current emissions trading, which takes place without there being a clear budget to ensure that overall emissions decline, and which also fudges the crucial issue of who owns the atmosphere.) With carbon permits – which will increase in value as they gradually decline in numbers to meet the global contraction budget – becoming prized property, there will be strong incentives for efficiency and the rapid uptake of clean energy technologies.

So whilst tackling global warming, 'contraction and convergence' would also go a long way towards reducing the appalling inequalities of today's world. Nor need it usurp the Kyoto Protocol: it could instead become a logical extension to the climate negotiations once the Kyoto 'first commitment period' mandate runs out in 2012.

I am convinced that 'contraction and convergence' provides the only solution to the problem of global warming which is both workable and logical, and which establishes a clear framework for deciding where we want to be in the future rather than simply relying on the guesswork of countless piecemeal measures. But in order for it to be accepted, governments first have to be persuaded to sign up to its provisions, something which can only be achieved with widespread popular support.

3. Stop all exploration and development of new oil, coal and gas

In the meantime, the world economy remains on fossil fuel autopilot. Most countries are already well above their permitted Kyoto emissions levels, and the International Energy Agency is pessimistic enough to predict world carbon dioxide emissions rising 70% above today's levels by 2030.[23]

New oil supplies are continually being opened up. The war fought in Iraq in spring 2003 almost certainly had US control of Middle Eastern oil supplies as one of its key – if unstated – objectives. Iraq has the second largest reserves in the world, and Saudi Arabia the largest.

The oil company BP is building a pipeline through Turkey and Georgia to take new crude supplies from Azerbaijan out to the Mediterranean. Exxon-Mobil, consistently the oil company with the most retrograde stance on global warming, announced in May 2003 the largest quarterly profits in corporate history: in the previous few weeks it had earned £2.2 million an hour.[24] Notwithstanding everything we know about climate change, oil is still a boom industry.

This development path is not just illogical, it is bordering on the suicidal. There are already enough oil supplies in 'proven reserves' to utterly destabilise the world's climate, and yet every single day thousands of highly-skilled petroleum geologists and other experts struggle to find new reserves. The environmental group Greenpeace worked out as long ago as 1997 that burning anything more than a quarter of existing fossil fuel reserves would lead to dangerous rates of global warming.[25] The conclusion to be drawn from this 'carbon logic' is obvious: that there should be a worldwide halt to the exploration and development of new oil, coal and gas reserves, because even existing reserves should never be burned as fuel.

This last point has become the rallying cry for an incipient movement seeking to combat global warming. Linking indigenous struggles against oil pollution and military repression as far afield as Ecuador and Cameroon with anti-corporate campaigners in the UK and US, this movement is sowing the seeds for a clean energy future. Their methods involve lobbying, petitioning and direct action, and targets range from governments – whose export agencies and overseas aid budgets still pour millions into new fossil fuel developments every year – to international financial institutions like the World Bank and the oil giants themselves.

Exxon (Esso in the UK) is currently also the target of a vociferous worldwide boycott campaign trying to change the company's head-in-the-sand attitude to climate change. But like Kyoto, this campaign can only be a first step: we have to stop the problem being made worse before we can begin to implement genuine solutions. We have to turn the fossil fuels juggernaut (or oil tanker) around, and the growing movement against new fossil fuels is surely the best way to start.

4. Take personal action to reduce emissions

It's easy – and in many ways justified – to blame Bush and the oil companies for causing and perpetuating climate change. But we all have a role to play in reducing emissions by altering the way we live our own lives. Current UK per capita emissions total 9.6 tonnes of carbon dioxide a year, whereas a 'sustainable' per person quota has been calculated at 2.45 tonnes.[26] In other words we Western consumers on average need to reduce our emissions to less than a quarter of their current levels.

It sounds impossible, but relatively painless (and often beneficial) shifts in lifestyle can make this target more than attainable for the average British person. More than half of the UK's current emissions come from sectors of the economy where consumers play a direct role, from cars to kettles.

Space heating consumes half of domestic energy use – and can be easily reduced by extra insulation, double-glazed windows and a good-quality gas boiler. A draughty, badly-insulated Victorian house can take five times as much energy to heat as a fully-insulated modern house. Insulation is a low-cost, one-off investment (which may also pay for itself through lower bills) and a surprising amount of financial assistance is available through a variety of little-advertised government schemes, especially for those on a low income.

It's also now fairly cheap and easy for UK consumers to change to green electricity, which means you can use market forces to help expand the renewables sector at little more than the flick of a switch.

Transport is still the fastest-growing contributor to global warming. Anyone who drives their car more than 10,000 miles a year will probably have already used more than double their sustainable carbon budget. The alternatives here are simple: reduce car use, switch to foot or bicycle for shorter journeys, and buses or trains for longer ones.

For most city-dwellers, cars are an unnecessary luxury, and hinder the development and operation of public transport. Sometimes having access to a car *is* very useful: perhaps to carry heavy goods around, or for occasional travel into rural areas. I had to hire one to get to the floods in Wales in a hurry, but a cheaper option is to join a car-share scheme, which enables many people to pool their costs into one vehicle.

Aircraft are much, much worse. As I outlined in Chapter 1, my flight emissions during the research for this book blew my own carbon budget for about 20 years, and a single return flight to New York would likely be as much as the rest of an environmentally-conscious person's other emissions put together. The warming effects of the actual carbon emissions from a jet engine are tripled because of vapour contrails and the fact that pollutants are injected high up into the Earth's atmosphere, where they can do the most damage.[27]

Although it's difficult to find sustainable alternatives for essential intercontinental travel, I can't see any need for internal flights around the UK and near-continental Europe, where far less polluting train services can serve just as well. I have boycotted these for several years, and have often enjoyed far more rewarding journeys as a result. On a trip to Rome a few years ago, I relaxed on a rattly old sleeper train, enjoying a beer as we wound our way through glorious views of the Alps. Had I done the trip by plane, I would have spent most of the time in airport queues, and emerged in a bad mood hours later having missed some of Europe's best scenery.

Unfortunately a barrage of airline advertising seeks to obscure these realities, whilst I don't think I've ever seen a single advert singing the praises of continental train travel. More importantly, price incentives also work the other way round: it's far cheaper to fly from London to Edinburgh than to use the train, because the effects of pollution are not factored into the cost of a ticket, and air transport currently enjoys a whole multitude of hidden subsidies – including, remarkably, tax-free fuel. The air transport industry expends a huge amount of time and money lobbying government to ensure these subsidies are protected, and that passenger numbers continue to rise unchecked.

The recent proliferation of budget airlines has made this situation immeasurably worse, and the government in Britain is considering building new airports and runways at several sites throughout the southeast of England. This is the reverse of what needs to happen: demand management must be employed to reduce incentives to fly and to increase incentives for surface transport. Short-haul flights can and should be phased out altogether. As the old Chinese proverb goes: it's okay to take a step backward when you're standing on the edge of a cliff.

5. Keep repeating the climate change message

Don't leave climate change to the experts. When I began to get interested and campaign on this issue, I did so without having any formal scientific training, and was continually worried about being too ignorant to speak out. But climate change is a very simple issue, and everyone needs to get involved in combating it.

In the UK alone, millions of people belong to environmental and wildlife groups. If every one of them began to vote, boycott companies, complain to the media and generally get angry on the issue of global warming, we'd be halfway to a solution. It's only because of our silence that the carbon economy remains so powerful.

We live in a society consumed by denial. Politicians make the occasional speech about the gravity of the climate change crisis – Tony Blair among them – and then go straight back to business, commissioning more road- and airport-building, encouraging fossil fuel-intensive economic globalisation, and subsidising new oil pipelines and drilling rigs. As consumers, we claim to be worried about global warming – and many polls have confirmed this fear – but we still do remarkably little to change our own habits and lifestyles.

Just as Alaskan Eskimos worry about disappearing sea ice whilst simultaneously supporting further oil industry expansion, we all continue to live in ways which perpetuate and intensify the climate crisis. It's someone else's problem and we vaguely hope that someone else will sort it out.

It's not easy to confront these entrenched patterns of behaviour. For example, when a group of friends suggest we all go on holiday to somewhere like Thailand, and I sheepishly admit that I can't join them because of my fears about climate change, I feel like I'm acting obsessively, and fear voicing these concerns makes me sound like a fanatic and an eco-bore. But objectively speaking, I'm right: the flight to Thailand *would* have contributed many more tonnes of carbon dioxide to the atmosphere, and as well as worsening global warming it would have made me a hypocrite. Like most people, I don't want to be seen as a wild-eyed ranter carrying a sandwich board saying 'the end of the world is nigh' down Oxford Street. Yet that's exactly what I hear myself sounding like when I try to tell people about the magnitude of the global warming crisis.

These days climate change *does* occasionally make the media, although television in Britain has studiously ignored the issue, and at the moment seems to be operating a complete ban on environmental programming of any sort. (How else to make space for *Big Brother* and *I'm a Celebrity, Get Me Out of Here*?) A few articles make it into the papers about melting ice caps and dead penguins, but they share space with car adverts and hundreds more articles which reinforce current perceptions of 'reality'.

And this reality has no time at all for concerns about the wider impacts of the way we live: instead we are urged to buy more clothes, eat more exotic food, drive sportier cars, live vicariously through high-consuming celebrities like David Beckham, take more holidays in the sun, and ditch local food-producers for big supermarkets which supposedly offer dramatic price discounts.

The only way that this mindset will ever change is if more and more people stand up and speak the truth. It's a difficult truth, and an unpleasant one: that the world is facing climatic catastrophe, and we all have to change the way we live. But even just organising green electricity in your workplace, writing a letter to the local newspaper, cycling the kid to school and holidaying in Scotland rather than Australia can make a vital contribution. Don't leave it to someone else, and don't believe the 'reality' of mass denial, which is itself nothing more than a lie.

Don't be scared to speak out. I'm an eco-bore at parties, and you can be too. Get with the program!

PARTING WORDS

Time is of the essence. We don't have long to change the course of society before trends in greenhouse gas emissions begin to break through the 'safe-landing corridor' and condemn the planet to a hot and dangerous future.[28] I've often heard it said that people will only wake up once a monster storm hits Washington or a terrible flood cripples London. But once these things begin to happen it may already be too late. We have to act on what we know already, and time is quickly running out.

I ended my global journey more convinced than ever that climate change is a reality, and that its impacts have become undeniable. During my three years of travelling, despite all the complexities of the science and the different stories told by the people I met, I don't think I saw a single piece of counter-evidence. Nothing out there suggests that the planet is getting cooler or the climate more stable.

If there's one message above all that I want people to take from these pages, it's this: that all the impacts described here are just the first whispers of the hurricane of future climate change which is now bearing down on us. Like the canary in the coal mine, those who live closest to the land – the Eskimos in Alaska and the Pacific islanders – have been the first to notice. But they won't be alone for long. As I suspected when I first began to undertake this mission, the first signs are evident to anyone who chooses to look.

My journey around a warming world was the experience of a lifetime. I now have friends in five continents, from Alaska to Peru, and from China to Tuvalu. I shared drinks and talked long into the night with people whom I would never have dreamed of meeting before. And moreover, I can now put names and faces on the vague concerns I held before my departure. My abstract 'droughts', 'floods' and 'glaciers' are now people and places, each with their own personality, their own vivid beauty and wonder, and their own myriad reasons for existence.

No longer do any of the places I visited seem remote. And I hope your experience of sharing them with me has illustrated just a little of why it's so vitally important – for all of our sakes – that we act now together to tackle the biggest crisis that humankind has ever faced.

Epilogue: Six Degrees

One day, 251 million years ago, a giant volcanic eruption shook modern-day Siberia. Billions of tonnes of hot ash and gases poured into the atmosphere, sparking huge storms of acid rain. Once the clouds cleared, the sun shone hotter than ever before, and searing heat killed plants and animals where they stood all around the planet. The end-Permian mass extinction had begun.

It was the worst ever crisis to affect life on Earth, and by the end up to 95% of the world's species were dead. In the words of one expert, it was the time 'when life nearly died'.[1] Geologists examining the strata of rocks spanning the Permian-Triassic boundary were stunned to see an abundance of fossils suddenly giving way to monotonous black mudstone – a telltale sign of anoxia, the crippling shortage of oxygen, as billions of lifeless bodies decayed at the bottom of the sea, having washed down from the ravaged land masses.

This crisis wasn't caused by an asteroid, unlike the catastrophe that later wiped out the dinosaurs. It was caused by global warming. The Siberian volcanoes had released vast quantities of carbon dioxide from deep within the Earth's crust, thus warming the climate sufficiently for massive amounts of methane to 'burp' out of the ocean, triggering a runaway greenhouse effect.

Life in this 'post-apocalyptic greenhouse' was so severe that only one large land animal survived – the adaptable pig-like Lystrosaurus, which became the ancestor of the dinosaurs. It took another fifty million years for pre-extinction levels of biodiversity to return.

Geologists examining oxygen isotopes in the end-Permian rocks have recently put a figure on how much global warming was associated with this catastrophic mass extinction. That figure is six degrees Celsius.[2]

Skip forwards 251 million years to the present day. The world is warming fast, and the evidence is everywhere, from thawing glaciers to rising sea levels. In 2001 the IPCC released its landmark Third Assessment Report, which made projections for future warming over the next hundred years.

The upper limit was higher than in previous assessments.[3]

The scientists had raised it – to six degrees.

Afterword

It's now four years since I started the *High Tide* project, and looking back I sometimes wonder what has changed. The Kyoto Protocol, potentially the most momentous environmental agreement ever negotiated, is only now coming into force, nearly a decade after it was originally agreed. President Bush still swaggers around the White House, punctuating his frequent naps with occasional legislation to unravel America's last remaining environmental safeguards. There's no hurry – Bush and his handlers know that his second term will be sufficient to finish the job. At the Environmental Protection Agency, following the debacle of 2003's climate-censored Report on the Environment,[1] global warming is still the issue that dare not speak its name. US-based climate scientists are among the best in the world, but the Administration itself remains in total denial.

But global warming can't simply be wished away. The planetary alarm bells keep on ringing: it's almost as if the biosphere is trying to tell us something: 2003 tied with 2002 as joint-second warmest years ever[2] (1998 still tops the league, thanks to its strong El Niño). Pre-monsoon 2003 temperatures hit an oven-like 49°C in India,[3] while the European heatwave of the same summer killed upwards of 30,000 people. The soaring temperatures broke all known records – the extreme heat was eventually calculated as a 1 in 46,000-year event.[4] 'Statistically, this event should not have happened,' said Christoph Schär, of the Swiss Federal Institute of Technology, Zurich, with engaging scientific understatement.

Warnings from the wild come almost every day, from the mountain-tops to the ocean depths. In North America, the pikas – cute, furry, rabbit-like animals that live at high altitudes in the Rockies – are in decline because of the warming weather,[5] while in the Alps rare plant communities are being forced gradually upwards. Once they reach the peaks, the only way up will be 'to heaven', in the words of Austrian botanist Georg Grabherr.[6] In the seas, meanwhile, entire ecosystems are refashioning themselves in response to rising temperatures. Hundreds of thousands of UK seabirds failed to raise young during the 2004 breeding season due to a sudden collapse in numbers of sand eels, the birds' main food source.[7] Two worldwide studies published in 2003 found evidence of changes in wildlife behaviour ranging from earlier frog breeding and the arrival of migrant birds to a general move towards the poles for butterflies, fish and hundreds of other plant and animal species.[8]

One thing has certainly changed. When I began writing

High Tide, global warming was rarely discussed or referred to in the media. Now it makes headlines almost every day. In July 2004 the BBC ran a week-long series of reports from places already impacted by climate change, including *High Tide* locations Shishmaref and northern China.[9] As I sat in the makeshift studio in Kew Gardens waiting for my 20-second soundbite, I watched the reports flash up on the monitors, and reflected how climate change seemed to have suddenly gone mainstream. There were no sceptics interviewed to 'get the other side'. The BBC reporters treated the realities of global warming, at long last, as solid fact. A month earlier, indeed, a Sky Television climate debate I was booked to appear on fell through at the last minute when the researcher was unable to find anyone who was prepared to oppose my position. Only professional contrarians and the odd maverick still deny the obvious.

I've tried to stay in touch with the people that I met on my travels, and can give a few updates here for interested readers. In Alaska, Shishmaref continues to erode into the sea. Robert Iyatunguk e-mailed me recently with news. 'We had a big one last fall,' he reported. 'Water was coming over the bluffs, the roads, we lost a lot of ground. With very high tides, the storm lasted for two days.' A subsequent e-mail raised the question of whether the village might seek UN funding, given the lack of interest from the US federal government. 'We natives are being put aside or neglected,' Robert complained. 'We are losing a lot of land. I don't think we can stand nature's wrath much longer.'

In Tuvalu, high tides again engulfed much of Funafuti

in February 2004. According to Agence France-Presse, 'the tides flooded homes, offices and even part of the airport'. The AFP reporter described, much as I saw, 'frightening springs of seawater' bubbling up through the ground across the island.[10] Unfortunately for those wanting to leave, the country's much-vaunted 'migration programme' has been beset with procedural difficulties. Because of new restrictions imposed by the New Zealand government, the latest information I have is that the supposed quota of 75 people a year has yet to be met.

It's even more difficult to find out what's going on in northern China. Reuters reported in March 2004 that one of the 'worst dust storms to hit northern China in years' had blanketed Beijing and turned skies in Inner Mongolia from blue to red and then black.[11] I have no way of getting word from Yang Pangon village about their battle with the advancing sands, or of finding out what has happened to Ye Yinxin, the sole remaining resident of the ghost village next to the former Qingtu Lake. I can only hope that if Mr and Mrs Dong are also forced to become refugees, they are met with as much kindness, warmth and hospitality as they showed to me when I stayed with them.

Peru is also suffering water shortages. In March 2004, Lima's SEDAPAL water authority announced the imminent start of water rationing in the capital.[12] Although this measure wasn't specifically linked to the disappearing glaciers, there can be little doubt that the reducing storage capacity of Andean ice fields will leave Peruvians with much less leeway in future times of drought. Indeed, the president of Peru's National Environment Council, Patricia Iturregui, warned in July 2004 that 'if climatic conditions remain as they are, all the glaciers in Peru below 5,500m will disappear by around 2015'.[13] This pessimistic prediction spells doom for the Sullcon glacier I visited, which supplies Lima's River Rimac and whose highest point reaches only 5,650 metres. I've also had an e-mail from a reader who retraced my steps up the valley to Yuracmayo (though he wasn't able to reach the glacier, due to the same altitude sickness problem that I suffered) and reports that the village leader, who has lived in the area all his life, 'could clearly provide witness about the diminishing snow and glaciers in the area'.[14]

It's been a busy year for my friends at HurricaneTrack.com: the last I heard from Mark Sudduth, he and the whole team were stuck in Florida, having run out of gas in the flooded aftermath of Hurricane Frances. The team made it into the eye of Hurricane Charley, a smaller storm that exploded to Category 4 just before making landfall in south-west Florida. Having left a trail of destruction across the state, Charley is set to become the second most expensive hurricane ever to hit the US after 1992's Hurricane Andrew. Ironically, just as Frances barrelled towards Florida's east coast, President Bush was standing up to make his keynote address at the Republican National Convention. No mention there of the fact that August 2004 was the busiest ever for Atlantic tropical cyclone activity. As I make clear in my chapter on hurricanes, it's no simple matter to make a link between increased tropical storm activity and global warming, but a graph on the US Weather Service's Climate Prediction Center website[15] shows a clear and increasing trend towards higher sea surface temperatures – and warm water is fuel to hurricanes just as uranium is to nuclear bombs.

Most unusual of all was the sudden appearance in March 2004 of a hurricane-like formation in the South Atlantic – an event any tropical forecaster worth his or her salt could tell you is not supposed to happen. The storm slammed into Brazil's east coast, causing up to a dozen fatalities. For an expert view I e-mailed *High Tide* interviewee Hugh Willoughby, who has now left the NOAA's Hurricane Research Division for a new post at the International Hurricane Research Center of Florida International University.

'Ah yes, "Hurricane" Catarina,' came the reply. 'My view is that it was a hurricane, although there is intense debate among the tropical cyclone crowd.' And could global warming have a role in its unannounced appearance? Willoughby's answer was a resounding 'perhaps'. Either way, hurricane monitoring services may now have to be extended to an entirely new ocean basin.

I rather rashly promised to answer 'all e-mails' in the first edition of *High Tide*, and have had to dedicate an increasingly large amount of time to fulfilling this pledge. But despite the heavy time commitment, it remains both rewarding and fascinating to hear from readers all over the world what they think about the book, as well as reports about their own experiences of global warming from places as far afield as Sweden, Australia and Cambodia. However,

I've tried to shift as much as possible of the discussion to my website, www.marklynas.org, so everyone can join in, and I also post near-daily blog articles about new global warming-related issues that crop up around the world.

One final word. *High Tide* was never meant to be a 'normal' book. Although its obvious task is to prove that global warming is real and already underway, this particular battle is nowadays pretty much won, as the dwindling band of sceptics proves. In any case, proving global warming was only ever the means to an end. The end itself, *High Tide*'s underlying mission, remains the same: to challenge all of us to face up to the implications of this reality, myself included. Here there's much less cause for celebration. In truth, we're still losing. Europe is increasingly divided. Kyoto is a shadow of what it was originally intended to be. Worst of all another Bush term is underway. What's still lacking most of all is a global citizens' movement demanding that governments seriously confront the threat of climate change. And it's this new citizens' movement that I hope readers of *High Tide* will play a part in forging.

APPENDIX:

Campaigns and Contacts

Contact the author at www.marklynas.org. I try to answer all emails, and would be happy to help readers who are looking for some way to get involved. I'll be updating the website continuously as this burgeoning movement evolves – don't miss my 'blog' for the latest news! I'll also post updates in other relevant areas covered by this book, so check in regularly. Here are some good starting points, however:

CAMPAIGNS

Global Commons Institute
www.gci.org.uk
GCI is the home of Contraction and Convergence, and the best first place to get involved in advocacy on its behalf.

Friends of the Earth
www.foe.co.uk
With national campaigns and many local groups, a good place to start tackling climate change locally.

Greenpeace
www.greenpeace.org
Another consistent campaigner against climate change.

WWF

www.panda.org

Worldwide campaigning organisation which works on climate as well as other environmental and conservation issues.

Rising Tide

www.risingtide.org.uk

Network of radical grassroots groups campaigning against global warming.

People and Planet

www.peopleandplanet.org

UK-based student campaigning network, with climate change as a top priority.

StopEsso

www.stopesso.com

International campaign against the corporate world's most retrograde climate denier.

Centre for Science and Environment

www.cseindia.org

Delhi-based organisation campaigning on climate, promoting general environmental awareness, and producing the excellent fortnightly 'Down to Earth' magazine.

Sierra Club

www.sierraclub.org

America's oldest environmental organisation, with climate change as a key focus.

National Resources Defense Council

www.nrdc.org

US-based environmental NGO also with a strong focus on global warming.

National Environmental Trust
www.environet.policy.net
Outspoken campaigning group which focuses some welcome vitriol on
the Bush Administration.

Climate Action Network
www.climatenetwork.org
Wide coalition of groups campaigning on climate change, with 11 offices
covering different countries and regions.

Boycott Bush
www.boycottbush.net
Speaks for itself!

Kyoto Now!
www.kyotonow.org
Movement dedicated to the ratification of the Kyoto Protocol by the US
Senate.

Oilwatch
www.oilwatch.org.ec
Ecuador-based international network of Southern groups campaigning
against oil exploration and production, and in support of indigenous
peoples.

BOOKS

There are many other books and journal articles cited in the References
to this book, which should provide good sources of further information
for specialised research. For more generalised reading, see the list below
of my top eight climate books.

Intergovernmental Panel on Climate Change, *Climate Change 2001*
(Cambridge University Press, 2001).
The Third Assessment Report of the IPCC, published in three volumes.
The first deals with the scientific basis, the second with impacts,

adaptation and vulnerability, and the third mitigation. The absolute bedrock of any further research on climate change, and I've relied heavily on it throughout this book. It's also available on the web (see below).

Aubrey Meyer, *Contraction and Convergence: The Global Solution to Climate Change* (Schumacher Briefing No. 5, Green Books, 2000).
Essential reading for anyone interested in understanding the roots of the C&C idea, and how it might be taken forward.

Michael Benton, *When Life Nearly Died: The Greatest Mass Extinction of All Time* (Thames & Hudson, 2003).
Not a climate change book per se, but a salutary reminder of what happened in another epoch which suffered episodes of dramatic global warming.

John Houghton, *Global Warming: The Complete Briefing* (Cambridge University Press, 1997).
An essential scientific backgrounder for non-specialists, explaining things like the physics of atmospheric greenhouse gas forcing in an accessible, engaging way. A new edition was published in 2004.

Ross Gelbspan, *The Heat is On: The Climate Crisis, the Cover-up, the Prescription* (Perseus Publishing, 1998).
The classic climate change book for a US audience, now updated on the web (see below).

Tom Athanasiou and Paul Baer, *Dead Heat: Global Justice and Global Warming* (Seven Stories Press, 2002).
Contraction and convergence with a new political spin – that no other solution will be able to keep Southern emissions below tolerable limits.

Jeremy Leggett, *The Carbon War: Global Warming and the End of the Oil Era* (Penguin Press, 1999).
A journey through the international negotiations up to and including Kyoto, by the Greenpeace lobbyist who was central to them.

Guy Dauncey, *Stormy Weather: 101 Solutions to Climate Change* (New Society Publishers, 2001).

How everyone (and their dog) can get involved in helping prevent dangerous climate change, from eating organic food to organising car-share schemes.

RESOURCES

The Intergovernmental Panel on Climate Change
www.ipcc.ch

This has to be the first port of call for anyone interested in further information about climate change. The Summary for Policymakers has a terse rundown of the latest science from the 2001 IPCC Third Assessment Report, but the entire report is available on the web.

UN Framework Convention on Climate Change
www.unfccc.int

The UNFCCC Secretariat website, with full text of the Kyoto Protocol and Framework Convention, lots of background info, pdf versions of national communications, scans of climate change in the world media and lots more.

Worldwatch Institute
www.worldwatch.org

Worldwatch produces the annual State of the World Report, and Vital Signs, both of which contain valuable statistics and facts on the economic and ecological trends related to climate change.

OneWorld.net
www.oneworld.net

My former employer, and an invaluable and very comprehensive source of news and resources from right across the environment and development spectrum. Also has specific sections and a campaign on climate change.

World Resources Institute

www.wri.org

Another good source of rigorous and reliable information.

EcoEquity

www.ecoequity.org

Site run by Tom Athanasiou and Paul Baer of *Dead Heat* fame (see above), with further explorations of equity considerations within the climate crisis.

PlanetArk

www.planetark.org

Daily news service of environment-related Reuters news articles, with search facility.

ClimateArk

www.climateark.org

Climate change portal site, with web search facility, directories and latest headlines.

Heat is Online

www.heatisonline.org

Website run by Ross Gelbspan, with the latest news and a special section to keep track of the antics of climate sceptics.

Tyndall Centre

www.tyndall.ac.uk

The Centre brings together scientists, economists, social scientists and other experts to develop responses to climate change.

Climatic Research Unit, University of East Anglia

www.cru.uea.ac.uk

Another good resource for scientific information, with several very useful factsheets about everything from sea level rise to global temperature records.

Choose Climate – Flying off to a Warmer Climate?
www.chooseclimate.org/flying
Planning a flight? Not yet feeling guilty? Don't forget to check out your emissions first.

Carbon Calculator
www.safeclimate.net/calculator
Go on – calculate yours. It's better to know . . .

FILM

Spanner Films' documentaries *Baked Alaska* and *Going Under* – the TV versions of this book's Chapters 2 and 3 – are available to buy or watch online at www.spannerfilms.net

Notes

1: Britain's Wet Season

1   Morris, M. and Allison, R., 2000: '"Unheard of" rain swamps the south', *The Guardian*, 13 October 2000.

2   Morris, M., 2000: 'Alerts remain as clean-up bill nears £2 billion', *The Guardian*, 14 October 2000.

3   Marsh, T.J. and Dale, M., 2002: 'The UK floods of 2000/01 – a hydrometeorological appraisal'. *CIWEM Journal*, **16**, No.3, 180–188.

4   DEFRA, 2001: 'To what degree can the October/November 2000 flood events be attributed to climate change?' FD2304 Final Report, March 2001. http://www.defra.gov.uk/environ/fcd/floodingincidents/fd2304fr.pdf.

5   Marsh, T.J., 2001: 'The 2000/01 floods in the UK – a brief overview'. *Weather*, **56**, 343–345.

6   Osborn, T., Hulme, M., Jones, P. and Basnett, T., 2000: 'Observed trends in the daily intensity of United Kingdom precipitation'. *International Journal of Climatology*, **20**, 347–364. To conduct the study, Osborn and Hulme divided all wet-day rainfall amounts into one of ten categories, from drizzle (1) up to downpour (10). They were defined over the whole period (1961–95, using data recorded by 150 weather stations all over the British Isles), each category containing 10% of the total rain.

7   This is known as the Clausius-Clapeyron relationship.

8   Osborn, T.J. and Hulme, M., 2002: 'Evidence for trends in heavy
    rainfall events over the UK'. *Philosophical Transactions of the Royal
    Society series A*, **360**, 1313–1325.

9   Marsh, T.J., 2001: *Op. cit.* note 5. In southwest Scotland every single
    year in the 1990s has exceeded the long-term average for 'heavy'
    precipitation days: see Alexander, L.V. and Jones, P.D., 2001:
    'Updated precipitation series for the UK and discussion of recent
    extremes'. *Atmospheric Science Letters*, **1**, doi:10.1006/asle.2001.0025.

10 Karl, T.R. and Knight, R.W., 1998: 'Secular trends of precipitation
    amount, frequency, and intensity in the United States'. *Bulletin of the
    American Meteorological Society*, **79**, No.2, February 1998, 231–241.
    Data from 182 stations shows an increasing rainfall trend of 10%
    between 1910 and 1996, and over half the increase has come in the
    heaviest 10% of events. Moreover, there has been a 2% increase in
    the area affected by more frequent 'extremely heavy' (over 50 mm
    per day) precipitation events.

11 Frei, C. and Schar, C., 2001: 'Detection probability of trends in rare
    events: Theory and application to heavy precipitation in the Alpine
    Region'. *Journal of Climate*, **14**, 1568–1584. Swiss climatologists
    examined ninety-four years of rain gauge data from 113 weather
    stations in Switzerland's Alpine region. They concluded that for
    intense precipitation their analysis 'yielded substantial evidence of
    increasing trends in the course of this century'.

12 Hennessy, K.J., Suppiah, R. and Page, C.M., 1999: 'Australian
    rainfall changes, 1910–1995'. *Australian Meteorological Magazine*, **48**,
    1–13. In Australia, over the same 1910–95 time period, annual total
    rainfall rose 15% in New South Wales, South Australia, Victoria and
    the Northern Territories. Over the whole country there has been a
    10% increase in the average number of rain days. This has come
    partly through 10–45% increase in heavy rainfall over much of the
    country.

13 Milly, P., Wetherald, R., Dunne, K. and Delworth, T., 2002:
    'Increasing risk of great floods in a changing climate'. *Nature*, **415**, 31
    January 2002, 514–517.

14 Frich P. et al., 2002: 'Observed coherent changes in climate extremes
    during the second half of the 20th century'. *Climate Research*, **19**,
    No.3, 193–212. The study found that on a global scale, not only was

heavy rainfall on an upward trend, but that five-day precipitation totals (a good indicator of flooding) showed 'a general increase throughout large areas of the globe' – most notably in North America and western Russia. They also concluded that during the second half of the twentieth century the world had indeed become both warmer and wetter. 'Heavy rainfall events have become more frequent,' they confirmed, and 'wet spells produce significantly higher rainfall totals now than they did just a few decades ago'. The study concludes that 'these observed changes in climatic extremes are in keeping with expected changes under enhanced greenhouse conditions'.

15 Folland, C. et al., 2001: 'Observed Climate Variability and Change', Chapter 2 in IPCC, 2001: *Climate Change 2001: The Scientific Basis. Contribution of Working Group I to the Third Assessment Report of the Intergovernmental Panel on Climate Change*, Cambridge University Press, pp.158 and 142. The IPCC characterises this as an 'amplified response for the heavy and extreme events' in the statistical surveys, and finds that 'zonally averaged precipitation increased by between 7% to 12%' in the mid-latitudes during the twentieth century.

16 Environment Agency, 2001: 'Lessons learned: Autumn 2000 floods'. March 2001. http://www.environment-agency.gov.uk/commondata/ 105385/126637.

17 UK Meteorological Office, historic station data for Oxford. http://www.met-office.gov.uk/climate/uk/stationdata/ oxforddata.txt, viewed May 2003.

18 Hulme, M. et al., 2002: 'Climate Change Scenarios for the United Kingdom: The UKCIP02 Scientific Report'. *Tyndall Centre for Climate Change Research*, School of Environmental Sciences, University of East Anglia. p.49.

19 Bisgrove, R. and Hadley, P., 2002: 'Gardening in the Global Greenhouse: Impacts of Climate Change on Gardens in the UK: Summary Report', UK Climate Impacts Programme, Oxford. http://www.ukcip.org.uk/pdfs/Gardens%20report/ Summary%20report.pdf.

20 Bisgrove, R. and Hadley, P., 2002: *Ibid.*, p.11.

21 Sparks, T., Crick, H., Woiwood, I. and Beebee, T.: 'Climate change and phenology in the United Kingdom', pp.53–55 in Green, R.,

Harley, M., Spalding, M. and Zockler, C. (eds), undated (2001?): 'Impacts of climate change on wildlife', published by RSPB on behalf of English Nature, WWF-UK, UNEP World Conservation Monitoring Centre and the Royal Society for the Protection of Birds. http://www.english-nature.org.uk/pubs/publication/PDF/ ImpactsCChange.pdf.

22 Hulme, M. et al., 2002: *Op. cit.* note 18, p.10.

23 The calculation for this is as follows: For every 1°C rise in temperature the climatic zone moves 150 kilometres north. The 1990s were in central England 0.5°C warmer than the 1961–90 average (UKCIP report, see note 18, p.11), which is equivalent to a total shift north of the climatic zone of seventy-five kilometres. This equates to about twenty metres per year.

24 Hulme, M. et al., 2002: *Op. cit.* note 18, p.28.

25 The Woodland Trust, 2001: 'A Midsummer Night's Nightmare? The future of UK woodland in the face of climate change', http://www.woodlandtrust.org.uk/policy/publicationsmore/ climatechangereport.pdf, p.7.

26 Bisgrove, R. and Hadley, P., 2002: *Op. cit.* note 19, p.8.

27 The Woodland Trust, 2001: *Op. cit.* note 25, p.12. This conclusion is necessarily speculative because it is based on what we know about these species today – no one can be absolutely sure how they may adapt to a globally-warmed future.

28 Hill, J., Thomas, C. and Huntley, B.: 'Impacts of climate and habitat availability on range changes in the Speckled Wood butterfly', pp.13–15 in Green, R., Harley, M., Spalding, M. and Zockler, C. (eds), undated (2001?): 'Impacts of climate change on wildlife', *op. cit.* note 21.

29 Harrison, P., Berry, P. and Dawson, T. (eds), 2001: 'Climate Change and Nature Conservation in Britain and Ireland: MONARCH – Modelling Natural Resource Responses to Climate Change: Summary Report', UK Climate Impacts Programme, Oxford. http://www.ukcip.org.uk/pdfs/monarch/Summary_report.pdf, p.12.

30 Woodland Trust, *op. cit.* note 25, p.10. The Woodland Trust states that most trees spread across the UK by a rate of about one kilometre per year (i.e. one hundred kilometres per century) at the end of the last Ice Age. Assuming a temperature rise of 5°C by 2100,

trees would need to move north by 750 kilometres during the century (see note 23), 7.5 times faster than their natural capability.

31 'What if it's hot?', *RSPB Youth Pages,* http://www.rspb.org.uk/youth/environment/Features/whatifitshot.asp.

32 Harrison, P., Berry, P. and Dawson, T. (eds), 2001: *Op. cit.* note 29.

33 Historical car use and bicycle statistics from Department for Transport data, 2002: http://www.transtat.dft.gov.uk/tables/tsgb02/9/pdf/90702.pdf. Road traffic forecasts from Department for Transport, 'National Road Traffic Forecasts (Great Britain) 1997': http://www.roads.dft.gov.uk/roadnetwork/nrpd/heta2/nrtf97/nrtf05.htm.

34 RAC, 2002: 'RAC Report on Motoring 2002', http://www.rac.co.uk/pdfs/report_2002.pdf.

35 Calculated using the sophisticated 'carbon calculator' on Future Forests' website. See http://www.futureforests.com/calculators/flightcalculatorshop.asp.

36 IPCC, 1999: 'Aviation and the Global Atmosphere: Summary for Policymakers', Cambridge University Press, http://www.ipcc.ch/pub/av(E).pdf. There are many uncertainties when considering the overall warming effect of aviation emissions, and the IPCC gives a range of two to five times the effect of CO_2 alone. See also Royal Commission on Environmental Pollution, 2002: 'The Environmental Effects of Civil Aircraft in Flight', London, http://www.rcep.org.uk/avreport.html, which arrives at a 'best estimate' of 2.7 times the effect of carbon dioxide alone.

37 Marshall, G., 2003: 'The carbon challenge – living for the future', *Clean Slate,* Journal of the Centre for Alternative Technology, Wales. Spring 2003 issue. This has been calculated at about 2.45 tonnes of carbon dioxide per person, per year, assuming that carbon entitlements are averaged out over the whole of the world's population, and that Earth systems absorb about half of current emissions. Incidentally, this is about a 75% reduction from current per capita British emissions.

38 DEFRA, 2001: 'The UK's third national communication under the United Nations Framework Convention on Climate Change', Department of Environment, Food and Rural Affairs, London, http://www.defra.gov.uk/environment/climatechange/3nc/pdf/

climate_3nc.pdf, pp.12–14. Transport is the biggest single energy user in the UK, accounting for 34% of final energy use in 1999. Households account for 29% (for which space heating accounts for about 55%, cooking about 5%, water heating about 25% and lights/appliances about 15%), industry, services and agriculture 37%.

39 DEFRA, 2001: *Ibid.*, p.12.

40 These statistics vary between winters and sources. These came from National Energy Action ('Campaigning for Warm Homes'), http://www.nea.org.uk/facts/wintermort.htm.

41 Hulme, M. et al., 2002: *Op. cit.* note 18, pp.61 & 65.

42 Hulme, M. et al., 2002: *Op. cit.* note 18, p.28.

2: Baked Alaska

1 This story is well known, and told in excellent detail by Dan O'Neill, in *The Firecracker Boys*, St Martin's Press, New York, 1994. This book's central theme is the amazing story of how the 'Father of the H-Bomb' Edward Teller almost achieved his plan to carve a new harbour out of the Alaskan coast by exploding six thermonuclear bombs.

2 Parson, E. et al., 2001: 'Potential Consequences of Climate Variability and Change for Alaska', Chapter 10 in National Assessment Synthesis Team, *Climate Change Impacts on the United States: The Potential Consequences of Climate Variability and Change*, Report for the US Global Change Research Program, Cambridge University Press, p.294. Available on http://www.usgcrp.gov/usgcrp/Library/nationalassessment/foundation.htm. The timing of break-up is recorded on the Tanana River to the exact minute, and the average date of spring thaw has advanced by about eight days.

3 Parson, E. et al., 2001, *ibid.*, p.292.

4 Weller, G. and Lange, M., 1999: *Impacts of Global Climate Change in the Arctic Regions.* Workshop on the Impacts of Global Change, 25–26 April 1999, Tromso, Norway. Published for the International Arctic Science Committee by the Center for Global Change and Arctic Systems Research, University of Alaska, Fairbanks.

5 Osterkamp, T. et al., 2000: 'Observations of Thermokarst and Its Impact on Boreal Forests in Alaska, USA', *Arctic, Antarctic, and Alpine Research*, **32**, 3, 303–315. See also: Osterkamp, T. and

Romanovsky, V., 1999: 'Evidence for Warming and Thawing of Discontinuous Permafrost in Alaska', *Permafrost and Periglacial Processes*, **10**, 17–37.

6   Parson, E. et al., 2001: *Op. cit.* note 2, p.293.

7   Alaska Natives Commission, *Final Report, Volume 1.* http://www.alaskool.org/resources/anc/anc_toc.htm.

8   Anisimov, O. and Fitzharris, B., 2001: 'Polar Regions (Arctic and Antarctic)' in IPCC 2001: *Climate Change 2001: Impacts, Adaptation, and Vulnerability. Contribution of Working Group II to the Third Assessment Report of the Intergovernmental Panel on Climate Change*, p.811.

9   The comparison with the Nile flow is made in Brown, L., 2000: 'Climate Change has World Skating on Thin Ice', *Worldwatch Issue Alert #7*, Worldwatch Institute, 29 August 2000.

10 Serreze, M. et al., 2000: 'Observational Evidence of Recent Change in the Northern High-Latitude Environment', *Climatic Change*, **46**, 159–207.

11 Magnuson, J. et al., 2000: 'Historical Trends in Lake and River Ice Cover in the Northern Hemisphere', *Science*, **289**, 1744–1746, 8 September 2000. This study looked at data from lakes and rivers in Canada, the US, Finland, Switzerland, Russia and Japan, and found that over the last century freezing and thawing dates had become later and earlier respectively by six days.

12 Arendt, A. et al., 2002: 'Rapid Wastage of Alaska Glaciers and Their Contribution to Rising Sea Level', *Science*, **297**, 382–386, 19 July 2002. This study used airborne laser altimetry to measure the thickness changes of sixty-seven Alaskan glaciers, and found them thinning on average by 0.5 metres a year. This yields about fifty-two cubic kilometres a year of meltwater, enough to raise global sea levels by 0.14 millimetres a year.

13 Anisimov, O. and Fitzharris, B., 2001: *Op. cit.* note 8.

14 Vaughan, D. et al., 2001: 'Devil in the Detail', *Science*, **293**, 1777–1779, 7 September 2001. This warming of the Antarctic Peninsula is exceptional for at least the last 500 years.

15 Dr David Vaughan, quoted in British Antarctic Survey press release, 19 March 2002: 'Satellite spies on doomed Antarctic ice shelf'.

16 Rothrock, D. et al., 1999: 'Thinning of the Arctic Sea-Ice Cover',

Geophysical Research Letters, **26**, 23, 3469–3472. This data comes from submarine cruises carried out between 1958–76 and 1993–97, which found a decline in Arctic sea ice thickness by 1.3 metres, or 40% during the earlier and later period. Another study looked at sea ice thickness data collected by the British submarine HMS *Sovereign* on a cruise across the Arctic in 1976 and compared it to a voyage undertaken by HMS *Trafalgar* along the same route twenty years earlier. The trend was a thickness decline of 43%. See Wadhams, P. and Davis, N., 2000: 'Further Evidence of Ice Thinning in the Arctic Ocean', *Geophysical Research Letters*, **29**, 24, 3973–3975.

17 See Parkinson, C. and Cavalieri, D., 2002: 'A Twenty-One year record of Arctic sea ice extents and their regional, seasonal and monthly variability and trends', *Annals of Glaciology*, **34**, 441–446. Wales is 20,761 square kilometres in area. 32,900 square kilometres of Northern Hemisphere sea ice is lost every year, about 3% per decade.

18 Parson, E. et al., 2001: *Op. cit.* note 2, p.292.

19 'Drying of Wetlands and Lakes in Alaska', in Land Cover Change Program – Alaska and Canada, undated. See http://picea.sel.uaf.edu/projects/lcluc2/lcluc2strat.html, this site has useful before and after satellite pictures of dried-up lakes and ponds, taken in 1985 and 1995. (Viewed 14 May 2003).

20 Anisimov, O. and Fitzharris, B., 2001: 'Polar Regions (Arctic and Antarctic)' in IPCC 2001: *Climate Change 2001: Impacts, Adaptation, and Vulnerability. Contribution of Working Group II to the Third Assessment Report of the Intergovernmental Panel on Climate Change*, p.823.

21 Parson, E. et al., 2001: *Op. cit.* note 2, p.296.

22 Wohlforth, C.: 'Spruce Bark Beetles and Climate Change', *Alaska Magazine*, March 2002.

23 BP Exploration and ARCO Alaska, 1997: 'Arctic Oil: Energy for Today and Tomorrow', July 1997.

24 Trustees for Alaska, 1998: *Under The Influence: Oil and the Industrialization of America's Arctic*. See Chapter IV, 'Oil Money Triumphant'.

25 For the latest assessment of Alaska Permanent Fund Corporation total value, see http://www.apfc.org. For a summary of annual

dividend payments, see http://www.pfd.state.ak.us/SUMMARY DIVAPPSPAYMENT.HTM.

26 US Fish and Wildlife Service, undated: *Arctic National Wildlife Refuge* leaflet. See also http://arctic.fws.gov.

27 See text of Alaska legislature, House Bill 117: http://www.legis.state.ak.us/basis/get_bill_text.asp?hsid=HB0117D& session=22.

28 'Petroleum People', *Alaska Oil & Gas Reporter*, 12 July 2001. http://www.oilandgasreporter.com/stories/071201/ peo_070101.shtml.

29 Ruskin, L., 2001: 'Toohey's appointment sparks cry of conflict in House', *Anchorage Daily News*, 19 June 2001.

30 'ANWR to Glacier Bay', Breakfast address by Cam Toohey, Special Assistant to the Secretary for Alaska, to the Alaska Resource Development Council, 2001. http://www.akrdc.org/membership/events/breakfast/toohey.html.

31 *The Arctic National Wildlife Refuge: Its People, Wildlife Resources, and Oil and Gas Potential*, published jointly in March 2001 by the Office of the Governor, the North Slope Borough and the Arctic Slope Regional Corporation.

32 Stirling, I. et al., 1999: 'Long-term Trends in the Population Ecology of Polar Bears in Western Hudson Bay in Relation to Climatic Change', *Arctic*, **52**, 3, 294–306.

33 Comiso, J., 2002: 'A rapidly declining perennial sea ice cover in the Arctic', *Geophysical Research Letters*, **29**, 20.

34 Parson, E. et al., 2001: O*p. cit.* note 2, p.301.

35 Parson, E. et al., 2001: O*p. cit.* note 2, p.289. Modelled temperature ranges vary from 4 to 10°C in the Canadian model to 3 to 6.5°C in the Hadley Centre model used by the US Assessment team.

36 Parson, E. et al., 2001: O*p. cit.* note 2, p.304.

37 Parson, E. et al., 2001: O*p. cit.* note 2, p.297.

3: Pacific Paradise Lost

1 Church, J.A. et al., 2001: 'Changes in Sea Level'. In: *Climate Change 2001: The Scientific Basis. Contribution of Working Group I to the Third Assessment Report of the Intergovernmental Panel on Climate Change*, p.664. The IPCC points out that analysis of the most recent satellite

data 'suggests a rate of sea level rise during the 1990s greater than the mean rate of rise for much of the twentieth century', although it is unsure about the precise causes.

2   Tuvaluans produce 0.465 tonnes of CO_2 per capita, according to their National Communication to the UNFCCC, http://unfccc.int/resource/docs/natc/tuvncl.pdf. This is obtained by dividing 4.65 gigagrammes of CO_2 by Tuvalu's 10,000 population to get this figure. For comparison, Australia produced 16.9 and Britain 9.5 tonnes of CO_2 per capita, in 1996 figures. Source: World Resources Institute, http://www.wri.org/wr-00-01/pdf/acln_2000.pdf.

3   Grimble, A., 1952: *A Pattern of Islands*, John Murray, London.

4   Noatia P. Teo, writing in *Tuvalu: A History*, published by the Institute of Pacific Studies and Extension Services, University of the South Pacific, Fiji, and the Ministry of Social Services, Tuvalu, 1983.

5   Macdonald, B., 1982: *Cinderellas of the Empire: Towards a history of Kiribati and Tuvalu*, Australia National University Press, Canberra.

6   Bryant, D. et al., 1998: *Reefs at Risk: A Map-Based Indicator of Threats to the World's Coral Reefs*, World Resources Institute, Washington DC, USA. http://www.wri.org/wri/pdf/reefs.pdf.

7   Ove Hoegh-Guldberg, a marine biologist at the University of Queensland, told me later in Australia that the 1998 El Nino was possibly 'the most serious impact on an ecosystem ever. We've never seen an ecosystem lose 16% of its key organism before. If we lost 16% of the rainforests in a single year, people would be screaming.'

8   Wilkinson, C. et al., 1999: 'Ecological and Socioeconomic Impacts of 1998 Coral Mortality in the Indian Ocean: An ENSO Impact and a Warning of Future Change?', *Ambio*, **28**, 2, 188–196.

9   Hoegh-Guldberg, O., 1999: 'Climate Change, Coral Bleaching and the Future of the World's Coral Reefs'. *Marine and Freshwater Research*, **50**, 839–66.

10 McLean, R.F. and Tsyban, A., 2001: 'Coastal Zones and Marine Ecosystems'. In *Climate Change 2001: Impacts, Adaptation and Vulnerability. Contribution of Working Group II to the Third Assessment Report of the Intergovernmental Panel on Climate Change.*

11 Leatherman, S., 2000: 'The Social and Economic Costs of Sea Level

Rise'. In Douglas, B. et al. (eds), *Sea Level Rise: History and Consequences*, Academic Press.

12 Information about 'managed realignment' is available from both the Department for Environment, Food and Rural Affairs and English Nature. For details of projects underway see http://www.english-nature.org.uk/livingwiththesea.

13 Church, J.A. et al., 2001: 'Changes in Sea Level'. In *Climate Change 2001: The Scientific Basis. Contribution of Working Group I to the Third Assessment Report of the Intergovernmental Panel on Climate Change*, p.642. The IPCC prediction is for somewhere between 9 centimetres and 88 centimetres over the next hundred years, and the central value of 48 centimetres 'gives an average rate of 2.2 to 4.4 times the rate over the twentieth century'.

14 That's 2.1 billion people. See Douglas, B., 2000: 'An Introduction to Sea Level'. In Douglas, B. et al. (eds), *Sea Level Rise: History and Consequences*, Academic Press.

15 Thirteen out of the world's largest twenty cities are on the coast. See Leatherman, S., 2000: O*p. cit.* note 11.

16 'Summary for Policymakers' in IPCC 2001: *Climate Change 2001: The Scientific Basis. Contribution of Working Group I to the Third Assessment Report of the Intergovernmental Panel on Climate Change*, Cambridge University Press, p.17.

17 Goldsmith, M. and Munro, D., 2002: *The Accidental Missionary: Tales of Elekana*, Macmillan Brown Centre of Pacific Studies, University of Canterbury, New Zealand.

18 NTF, 2002: 'Sea Level in Tuvalu: Its Present State'. National Tidal Facility, Australia. http://www.ntf.flinders.edu.au/TEXT/NEWS/tuvalu.pdf. Viewed March 2002, and since updated (see note 23).

19 Radio Australia, 16 October 2001: 'Scientific Evidence Does Not Support Tuvalu Sinking Theory'.

20 AFP, 28 March 2002: 'Global warming not sinking Tuvalu – But maybe its own people are'. http://203.97.34.63/tuvalu7.htm.

21 See http://www.john-daly.com, and also 'What's wrong with *Still waiting for Greenhouse*', an assessment by a group of concerned scientists: http://www.trump.net.au/~greenhou.

22 See http://www.cei.org/utils/printer.cfm?AID=3069, although an

article entitled 'Tuvalunacy', published in the *Washington Times* on 4 April 2002 is even more spectacularly biased (and wrong). See http://www.cei.org/gencon/019,02947.cfm, viewed 27 April 2003.

23 In NTF's defence, it should be noted that more recent analyses *have* noted a rise in sea level, albeit with a similarly short data span. Viewed on 27 April 2003, NTF had revised its 'Sea level in Tuvalu: its present state' document to read: 'nearly nine years of data show a rate of +0.9 millimetres per year'. See http://www.ntf.flinders.edu.au/TEXT/NEWS/tuvalu.pdf for the latest assessment.

24 This is also part of the 'Sea level in Tuvalu: its present state' analysis. *Op. cit.* note 18.

25 During El Niño events the usual east–west ocean currents go into reverse, allowing warm water to cross into the eastern Pacific. They also change sea levels dramatically – but temporarily – across the whole ocean. In 1998, for example, sea levels in March and April were 35 centimetres lower than normal due to El Niño, but had recovered by November.

26 Hunter, J., 2002: 'A Note on Relative Sea Level Change at Funafuti, Tuvalu'. Antarctic Cooperative Research Centre. http://www.antcrc.utas.edu.au/~johunter/tuvalu.pdf. Hunter's analysis shows an average sea level rise of 1 millimetre per year, which rises to 1.2 millimetres if other El Niño distortions are removed. This is well within the 1–2 millimetres per year IPCC estimate.

27 Hunter, J., 2002: personal communication.

4: Red Clouds in China

1 Yang, G., Xiao, H. and Tuo, W., 2001: 'Black windstorm in northwest China: a case study of the strong sand-dust storm on May 5 1993', Chapter 3 in Yang, Y., Squires, V. and Lu, Q. (eds), 2001: *Global Alarm: Dust and Sandstorms from the World's Drylands*, Asia Regional Coordinating Unit, Secretariat of the United Nations Convention to Combat Desertification, Bangkok, Thailand.

2 Sawin, J., 2003: 'Severe Weather Events on the Rise', in Worldwatch Institute, *Vital Signs 2003: The Trends That Are Shaping Our Future*, Worldwatch Institute, New York, p.92.

3   Qian, W. and Zhu, Y., 2001: 'Climate change in China from 1880 to 1998 and its impact on the environmental condition', *Climatic Change*, **50**, 419–444.

4   Kurtenbach, E., 2003: 'China says its north is facing water shortages and polluted', Associated Press, 6 June 2003.

5   CCICCD, 2000: *China National Report on the Implementation of United Nations Convention to Combat Desertification and National Action Programme to Combat Desertification*, Beijing, April 2000.

6   The precise contribution of direct land degradation remains unquantified. It also seems that rainfall in northwest China is linked to El Niño events – with El Niño decreasing and La Niña increasing precipitation totals. Since there were more frequent El Niños in the 1980s and 1990s, these may have contributed to the general drying trend. The extent to which rainfall fluctuations are part of inter-decadal and inter-centurial cycles also remains controversial. For more discussion on these issues see Wang, S. and Zhao, Z., 1981: 'Droughts and floods in China, 1470–1979,' in Wigley, T., Ingram, M. and Farmer, G. (eds), *Climate and History*, Cambridge University Press (UK), 1981; also Wang, S. et al., *Advances of Modern Climatology*, China Meteorological Press (China), p.265 [in Chinese].

7   IPCC 2001: *Climate Change 2001: Synthesis Report of the Third Assessment Report of the Intergovernmental Panel on Climate Change*, section 4.6, p.58 (pre-publication draft). The IPCC points out that 'A general drying of the mid-continental areas during summer is likely to lead to increases in summer droughts. This general drying is due to a combination of increased temperature and potential evaporation that is not balanced by increases in precipitation.'

8   Karoly, D. et al., 2003: 'Global warming contributes to Australia's worst drought', WWF Australia, Sydney, NSW, Australia. According to the authors, temperatures in the Murray-Darling river basin, the centre of the 2002 drought and usually one of Australia's most productive agricultural regions, were higher than in any earlier drought – soaring to 2°C above the previous average, and directly driving higher evaporation rates. These higher temperatures, the authors concluded, were not simply a natural aberration, but part of a longer-term trend of increasing warmth which was 'likely due to the increase in greenhouse gases in the atmosphere'.

9   Wetherald, R. and Manabe, S., 1999: 'Detectability of summer
    dryness caused by greenhouse warming', *Climatic Change*, **43**,
    495–511. Although the authors of the above study concluded that
    evidence of global warming-induced drying of continental interiors
    was unlikely to be 'definitive' before the middle of this century, the
    model predicted the largest reductions in semi-arid regions like
    southern North America, the Mediterranean and Central Asia, so it
    is possible these are already being observed.

10 Wang, S. et al., 2001: 'Twentieth-century climatic warming in China
    in the context of the Holocene', *The Holocene*, **11**, 3, 313–321. This
    assertion is based on data from tree rings and ice cores, which give a
    proxy temperature record going back 400 to 600 years.

11 Qian, W. and Zhu, Y., 2001: 'Climate change in China from 1880 to
    1998 and its impact on the environmental condition', *Climatic
    Change*, **50**, 419–444. This trend varies throughout the whole of the
    semi-arid northern region – from 1.8°C in some parts of Inner
    Mongolia to 0.5°C further west. Also see Zhao, Z., 1996: 'Climate
    Change and Sustainable Development in China's Semi-arid
    Regions', Chapter 4 in Ribot, J. et al. (eds), 1996: *Climate Variability,
    Climate Change and Social Vulnerability in the Semi-arid Tropics*,
    Cambridge University Press, Cambridge.

12 Wang, F. and Zhao, Z., 1995: 'Impact of climate change on natural
    vegetation in China and its implication for agriculture', *Journal of
    Biogeography*, **22**, 657–664.

13 Dai, A. and Trenberth, K., 1998: 'Global Variations in Droughts and
    Wet Spells: 1900–1995', *Geophysical Research Letters*, **25**, 17,
    3367–3370. Although the trends are small, and masked by variations
    like the El Niño cycle in the Pacific, even those droughts which are
    linked with El Niño now seem to be more severe than similar
    episodes in previous decades. 'These changes are qualitatively
    consistent with those expected from increased greenhouse gases in
    the atmosphere,' the authors conclude.

14 Waple, A. et al., 2002: 'Climate Assessment for 2001', *Bulletin of the
    American Meteorological Society*, **83**, 6, 938–938 (supplement).

15 Hoerling, M. and Kumar, A., 2003: 'The Perfect Ocean for Drought',
    Science, **299**, 691–694. Part of the ocean warming was due to the
    Pacific El Nino cycle – but not only were the higher ocean

temperatures 'unsurpassed during the twentieth century', they were also 'embedded within a multidecade warming trend' which was likely to be 'partly due to the ocean's response to increased greenhouse gases', according to the authors. Note that this mechanism would not necessarily affect Northern China, since China has a monsoon system like India rather than receiving most of its rainfall in winter cyclones, as does Central and West Asia. But Northern China's position at the semi-arid mid-latitudes is consistent with this scenario.

16 Zhai, P., Sun, A., Ren, F., Liu, X., Gao, B. and Zhang, Q., 1999: 'Changes of climate extremes in China', *Climatic Change*, **42**, 203–218. This survey found a 6% decrease in the number of rainy days per decade between 1951 and 1995, and a 3% decrease in annual precipitation overall.

17 Qian, W. and Zhu, Y., 2001: *Op. cit.* note 3.

18 Yang, G., Xiao, H. and Tuo, W., 2001: *Op cit* note 1.

19 Yang, Y. and Lu, Q., 2001: 'Dust-sandstorms: Inevitable consequences of desertification – a case study of desertification disasters in the Hexi Corridor, northwest China', Chapter 11 in Yang, Y., Squires, V. and Lu, Q. (eds), 2001: *Global Alarm: Dust and Sandstorms from the World's Drylands*, Asia Regional Coordinating Unit, Secretariat of the United Nations Convention to Combat Desertification, Bangkok, Thailand.

5: Hurricane USA

1 Elsner, J. and Kara, A.B., 1999: *Hurricanes of the North Atlantic: Climate and Society*, Oxford University Press.

2 Goldenberg, S., Landsea, C., Mestas-Nunez, A. and Gray, W., 2001: 'The Recent Increase in Atlantic Hurricane Activity: Causes and Implications', *Science*, **293**, 474–479.

3 Landsea, C., Nicholls, N., Gray, W. and Avila, L., 1996: 'Downward Trends in the Frequency of Intense Atlantic Hurricanes During the Past Five Decades', *Geophysical Research Letters*, **23**, 1697–1700.

4 Landsea, C., Bell, G., Gray, W. and Goldenberg, S., 1998: 'The Extremely Active 1995 Atlantic Hurricane Season: Environmental Conditions and Verification of Seasonal Forecasts', *Monthly Weather Review*, **126**, 1174–1193.

5   Goldenberg, S. et al., 2001: *Op. cit.* note 2.

6   Lander, M. and Guard, C., 1998: 'A Look at Global Tropical Cyclone
     Activity during 1995: Contrasting High Atlantic Activity with Low
     Activity in Other Basins', *Monthly Weather Review*, **126**, 1163–1173.

7   Chan, J. and Shi, J., 1996: 'Long-Term Trends and Interannual
     Variability in Tropical Cyclone Activity over the Western North
     Pacific', *Geophysical Research Letters*, **23**, 20, 2765–2767.

8   Nicholls, N., Landsea, C. and Gill, J., 1998: 'Recent Trends in
     Australian Region Tropical Cyclone Activity', *Meteorology and
     Atmospheric Physics*, **65**, 197–205.

9   According to the Concise Oxford English Dictionary (1990 edition)
     a manatee is 'any large aquatic plant-eating mammal of the genus
     Trichechus, with paddle-like forelimbs, no hind limbs, and a
     powerful tail'.

10   Davies, P., 2000: *Inside the Hurricane: Face to Face with Nature's
     Deadliest Storms*, Henry Holt and Company, New York.

11   NOAA press release: 'After Ten Years, Hurricane Andrew Gains
     Strength', issued 21 August 2002, http://www.nhc.noaa.gov/
     NOAA_pr_8-21-02.html.

12   Rappaport, E., 1993: 'Preliminary Report, Hurricane Andrew,
     16–28 August 1992', National Hurricane Center,
     http://www.nhc.noaa.gov/1992andrew.html.

13   Even a US Senate panel made this connection: see US Senate
     Bipartisan Task Force on Funding Disaster Relief, 1995: *Federal
     Disaster Assistance*, 104.

14   Pielke, R. Jr and Landsea, C., 1998: 'Normalized Hurricane Damages
     in the United States: 1925–95', *Weather and Forecasting*, **13**, 621–631.

15   Goldenberg, S. et al., 2001: *Op. cit.* note 2.

16   Associated Press, 2001: 'Lili Leaves Soggy Mark on Louisiana', 3
     October 2001.

17   Knutson, T. and Tuleya, R., 1999: 'Increased Hurricane Intensities
     with CO_2-Induced Warming as Simulated using the GFDL
     Hurricane Prediction System', *Climate Dynamics*, **15**, 503–519.

18   Knutson, T., Tuleya, R., Shen, W. and Ginis, I., 2001: 'Impact of
     CO_2-Induced Warming on Hurricane Intensities as Simulated in a
     Hurricane Model with Ocean Coupling', *Journal of Climate*, **14**, 11,
     2458–2468.

19 Guiney, J. and Lawrence, M., 1999: 'Preliminary Report, Hurricane Mitch, 22 October – 5 November 1998', National Hurricane Center, http://www.nhc.noaa.gov/1998mitch.html.

20 Pan-American Health Organisation, undated: 'El Huracán Mitch en Honduras', http://www.paho.org/Spanish/PED/gm-honduras.pdf.

21 Associated Press, 1999: 'A Look Back at Mitch's Rampage', 8 June 1999.

22 Associated Press, 1999: 'Central Americans Relive Horror of Mitch', 23 May 1999.

23 International Federation of Red Cross and Red Crescent Societies, 1999: *World Disasters Report 1999*.

24 International Federation of Red Cross and Red Crescent Societies, 2000: *World Disasters Report 2000*.

25 Joint Typhoon Warning Center, 1991: 'Tropical Cyclone 02B', http://www.npmoc.navy.mil/jtwc/atcr/1991atcr/pdf/nio/02b.pdf.

26 Agence France-Presse, 1991: 'Bangladesh tidal wave: Clinging to life as the skies split', *The Independent*, 5 May 1991.

27 Allen-Mills, T., 1991: 'Deluge worsens misery in ravaged Bangladesh', *Sunday Times*, 12 May 1991.

28 UN Centre for Regional Development, 1991: 'Cyclone damage in Bangladesh: Report on field study and investigations on the damage caused by the cyclone in Bangladesh in 29–30 April 1991', United Nations Centre for Regional Development, Nagoya, Japan, December 1991.

29 Waldron, K., 1991: 'Delta awash with 125,000 dead; Karl Waldron in Dhaka tells of a despair so great that Bangladesh cannot begin to ease the agony', *The Independent*, 5 May 1991.

30 UN Centre for Regional Development, 1991: *Op. cit.* note 28.

31 Lou's Weather Watch Interview Page, 10 March 2001: 'Interview with Stacy Stewart of the National Hurricane Center', http://community-2.webtv.net/ltursi/LOUSWEATHER WATCH0/page3.html.

32 Joint Typhoon Warning Center, 1999: 'Tropical Cyclone 05B', http://www.npmoc.navy.mil/jtwc/atcr/1999atcr/pdf/05b.pdf.

6: Peru's Melting Point

1 IPCC 2001: *Climate Change 2001: Impacts, Adaptation, and Vulnerability. Contribution of Working Group II to the Third Assessment Report of the Intergovernmental Panel on Climate Change*, p.208.

2 Ames, A., 1998: 'A documentation of glacier tongue variations and lake development in the Cordillera Blanca, Peru', *Zeitschrift fur Gletscherkunde und Glazialgeologie*, **34**, 1, 1–36.

3 Ames, A., 1998: *Ibid.*

4 Morales Arnao, B., 2000: 'Los eternos nevados en el Perú están retrociendo en forma cada vez más accelerada', Chapter 2 in *El Medio Ambiente en el Perú – Año 2000*, Instituto Cuánto, Lima, p.19.

5 Morales Arnao, B., 2000: *Ibid.*

6 Morales Arnao, B., 2000: *Op. cit.* note 4, p.20, 'Cuadro N.2.2: Inventario de glaciares – 18 cordilleras'.

7 Morales Arnao, B., 2000: *Ibid.*

8 Kaser, G., 1999: 'A review of the modern fluctuations of tropical glaciers', *Global and Planetary Change*, **22**, 93–103.

9 Thompson, L. et al., 2002: 'Kilimanjaro Ice Core Records: Evidence of Holocene Climate Change in Tropical Africa', *Science*, **298**, 589–593.

10 Kaser, G., 1999: *Op. cit.* note 8.

11 Rivera, A. et al., 2002: 'Use of remotely sensed and field data to estimate the contribution of Chilean glaciers to eustatic sea-level rise', *Annals of Glaciology*, **34**, 367–372.

12 Fagre, D., 2001: 'Glacier monitoring in Glacier National Park', U.S. Geological Survey, see http://nrmsc.usgs.gov/research/glaciers.htm.

13 Arendt, A. et al., 2002: 'Rapid Wastage of Alaska Glaciers and Their Contribution to Rising Sea Level', *Science*, **297**, 382–386, 19 July 2002.

14 Haeberli, W. and Hoelzle, M., 1995: 'Application of inventory data for estimating characteristics of and regional climate-change effects on mountain glaciers: a pilot study with the European Alps', *Annals of Glaciology*, **21**, 206–212.

15 Kääb, A. et al., 2002: 'The new remote-sensing-derived Swiss glacier inventory: II. First results', *Annals of Glaciology*, 34, 362–366.

16 Haeberli, W. and Hoelzle, M., 1995: *Op. cit.* note 14.

17 Waddington, R., 2003: 'Excessive heat takes toll on Swiss Alpine glaciers', *Reuters*, 25 August 2003.

18 Qin, D. et al., 2000: 'Evidence for recent climate change from ice cores in the central Himalaya', *Annals of Glaciology*, 31, 153–158.

19 Fujita, K. et al., 1997: 'Changes in glaciers in Hidden Valley, Mukut Himal, Nepal Himalayas, from 1974–1994', *Journal of Glaciology*, 43, 145, 583–588.

20 Lake Geneva's volume is 90,000,000,000 cubic metres, or ninety cubic kilometres. Data from the International Lake Environment Committee, http://www.ilec.or.jp/database/eur/dseur071.html.

21 Dyurgerov, M. and Meier, M., 2000: 'Twentieth-century climate change: evidence from small glaciers', *Proceedings of National Academy of Sciences*, USA, 97, 4, 1406–1411.

22 Dyurgerov, M., 2002: 'Glacier Mass Balance and Regime: Data of Measurements and Analysis', Institute of Arctic and Alpine Research, University of Colorado, Boulder. Occasional paper, 275 pp. http://instaar.colorado.edu/other/download/OP55_glaciers.pdf.

23 Haeberli, W. et al., 1999: 'On rates and acceleration trends of global glacier mass changes', *Geografiska Annaler*, 81A, 585–591.

24 Peru – Mountain Ecosystems World Meeting: 'Mountains towards 2020: Water, Life and Production', Huaraz, 12–14 June 2002.

25 The Spanish acronym INAGGA stands for 'Instituto Andino de Glaciología y Geoambiente', see http://www.itete.com.pe/inagga.

26 International Year of Mountains Coordination Unit, 2002: 'Water', factsheet on International Year of Mountains website, see http://www.mountains2002.org/i-water.html.

27 Noriega Pissani, R., 2000: 'Lima apunta a la cuenca del Mantaro para abastecerse de agua en el siglo XXI', Chapter 6 in *El Medio Ambiente en el Perú – Año 2000*, Instituto Cuánto, Lima.

28 Kaser, G. et al., 2003: 'The impact of glaciers on the runoff and the reconstruction of mass balance history from hydrological data in the tropical Cordillera Blanca, Perú', *Journal of Hydrology*, in press.

29 Noriega Pissani, R., 2000: *Op. cit.* note 27.

30 Morales Arnao, B., 2000: *Op. cit.* note 4.

31 Bernard Pouyaud, 2002, personal communication.

32 Ramirez, E. et al., 2001: 'Small glaciers disappearing in the Tropical Andes. A case study in Bolivia: the Chacaltaya glacier, 16°S', *Journal of Glaciology*, **47**, 157, 187–194.

33 Kargel, J. et al., 2002: 'A World of Changing Glaciers: Hazards, Opportunities, and Measures of Global Climate Change', *Eos, Transactions*, American Geophysical Union, **83**, 19, Spring Meeting Supplement, Abstract U31A–04.

34 International Year of Mountains Coordination Unit, 2002: *Op. cit.* note 26.

35 Kulkarni, A. and Bahuguna, I., 2002: 'Glacial retreat in the Baspa basin, Himalaya, monitored with satellite stereo data', *Journal of Glaciology*, **48**, 160, Correspondence 171–172.

36 NASA News Release, 'Decline of world's glaciers expected to have global impacts over this century', 29 May 2002. http://www.gsfc.nasa.gov/news-release/releases/2002/02–76.htm.

37 Kargel, J., 2003: personal communication.

38 Morales Arnao, B., 1999: 'Estudios de vulnerabilidad de recursos hídricos de alta montaña en el Perú', in *Perú: Vulnerabilidad Frente al Cambio Climático*, Consejo Nacional de Ambiente, Lima. p.209.

39 Morales Arnao, B., 2000: *Op. Cit.* note 4. Ice area in the Cordillera Central fell from 116.6 to seventy-nine square kilometres between 1970 and 1997. The volume of Lake Windermere is about 0.3 billion cubic metres (or 0.3 cubic kilometres) according to World Resources Institute, http://www.wri.org/wri/wr-98-99/pdf/wr98_wa3.pdf. The International Lake Environment Committee gives a similar volume of 315,000,000 cubic metres. See http://www.ilec.or.jp/database/eur/deur11.html.

7: Feeling the Heat

1 Prentice, I. et al., 2001: 'The Carbon Cycle and Atmospheric Carbon Dioxide' in IPCC 2001: *Climate Change 2001: The Scientific Basis. Contribution of Working Group I to the Third Assessment Report of the Intergovernmental Panel on Climate Change*, Cambridge University Press, p.185.

2 'Summary for Policymakers' in IPCC 2001: *Climate Change 2001*, Cambridge University Press.

3   Waple, A. et al., 2002: 'Climate Assessment for 2001', *Bulletin of the American Meteorological Society*, **83**, 6, 938–938 (supplement).

4   'Summary for Policymakers' in IPCC 2001: *Climate Change 2001*, Cambridge University Press.

5   IPCC 2001: *Climate Change 2001: Synthesis Report of the Third Assessment Report of the Intergovernmental Panel on Climate Change*, section p.48 (pre-publication draft).

6   These methane hydrates total about 10,000 billion tonnes of carbon, or more than double the world's entire fossil fuel reserves. See Suess, E. et al., 1999: 'Flammable Ice', *Scientific American*, November 1999. There is strong evidence that a methane 'burp' into the atmosphere was behind the high temperatures experienced during the Late Palaeocene Thermal Maximum, 55 million years ago, when rates of carbon addition to the atmosphere were directly comparable to the rate of addition today due to fossil fuel combustion. See Katz, M. et al., 1999: 'The Source and Fate of Massive Carbon Input During the Latest Paleocene Thermal Maximum', *Science*, **286**, 1531–1533.

7   Cox, P. et al., 2000: 'Acceleration of global warming due to carbon-cycle feedbacks in a coupled climate model', *Nature*, **408**, 184–187. Adding this extra warming to the worst-case scenario arrives at a total of $7.3\,^\circ C$.

8   Jones, C. et al., 2003: 'Strong carbon cycle feedbacks in a climate model with interactive CO_2 and sulphate aerosols', *Geophysical Research Letters*, **30**, 9, 1479.

9   Pearce, F., 2003: 'Heat will soar as haze fades', *New Scientist*, 7 June 2003.

10 Leggett, J., 1999: *The Carbon War: Global Warming and the End of the Oil Era*, Penguin Books, London.

11 Natural Resources Defense Council, 2001: 'Slower, Costlier and Dirtier: A Critique of the Bush Energy Plan', May 2001. http://www.nrdc.org/air/energy/scd/scd.pdf.

12 Natural Resources Defense Council, 2001: *Ibid.*

13 Industry enjoyed 714 contacts with Cheney's task force, whilst non-industry representatives, including environmental groups, had only 29. NRDC press release, 21 May 2002: 'Data Shows Industry had Extensive Access to Cheney's Energy Task Force', http://www.nrdc.org/media/pressreleases/020521.asp.

14 Figures from the Center for Responsive Politics website, viewed February 2003. http://www.opensecrets.org/bush/index.asp.

15 Kay, K., 2001: 'Analysis: Oil and the Bush Cabinet', *BBC News Online*, 29 January 2001. *http://news.bbc.co.uk/1/hi/world/americas/1138009.stm.*

16 Mayrack, B., 2001: 'Bush's New Chief of Staff Once Fought for Polluters', *The Public i*, Center for Public Integrity. http://www.publicintegrity.org/dtaweb/report.asp?ReportID=137&L1=10&L2=70&L3=15&L4=0&L5=0& State=&Year=2001.

17 Prescott, J., 2001: 'Children pay for political errors', *The Observer*, 1 April 2001.

18 Sohlman, E., 2001: 'EU: No Trade Retaliation Due to US Kyoto Refusal', *Reuters*, 31 March 2001.

19 AFP, 2001: 'US feels heat after dumping global warming treaty', 29 March 2001.

20 CNN.com, 2001: 'EU to press US over Kyoto', 1 April 2001. http://www.cnn.com/2001/WORLD/europe/italy/04/01/kyoto.eu.

21 ECO NGO newsletter, 20 July 2001.

22 Global Commons Institute, 2003: *The Essential of Contraction and Convergence*, GCI, London. http://www.gci.org.uk/refs/UNEPFI6.pdf.

23 IEA, 2003: *World Energy Outlook*, http://www.worldenergy outlook.org/weo/papers/Weoc02.pdf.

24 Macalister, T., 2003: 'War propels Exxon profits to record £7 billion', *The Guardian*, 2 May 2003.

25 Hare, B., 1997: *Fossil Fuels and Climate Protection: The Carbon Logic*, Greenpeace International, Amsterdam.

26 Marshall, G., 2003: 'The carbon challenge – living for the future', *Clean Slate*, Journal of the Centre for Alternative Technology, Wales. Spring 2003.

27 IPCC, 1999: 'Aviation and the Global Atmosphere: Summary for Policymakers', Cambridge University Press, http://www.ipcc.ch/pub/av(E).pdf. See also Royal Commission on Environmental Pollution, 2002: 'The Environmental Effects of Civil Aircraft in Flight', London, http://www.rcep.org.uk/avreport.html.

28 Athanasiou, T. and Baer, P., 2002: *Dead Heat: Global Justice and Global Warming*, Seven Stories Press, New York. This book has

extensive discussion on what a 'safe-landing corridor' might be, and how long we have to stay within it.

Epilogue

1   Benton, M., 2003: *When Life Nearly Died: The greatest mass extinction of all time*, Thames & Hudson, London, 2003. This book tells the story in fascinating detail.

2   Benton, M., 2003: *Ibid.*, pp.267–277.

3   The IPCC upper end projection is actually 5.8 degrees Celsius, but it doesn't include some of the more extreme feedbacks like 'methane burps' or Amazonian drying. The latest models, which use an integrated 'Earth systems' approach, tend to revise the estimates of climate sensitivity upwards. See Jones, D. et al., 2003: 'Strong carbon cycle feedbacks in a climate model with interactive CO_2 and sulphate aerosols', *Geophysical Research Letters*, **30**, 1479–1482. In addition, the end-Permian temperature rise probably took place over a much longer period of several thousand years, so the warming during this century could be even more disastrous for life. However, there are important differences between then and now, for instance the distribution of the continents, and the effect of acid rain on land-based life. For much more on this, and an interesting parallel with the current 'sixth mass extinction' caused mostly by habitat loss and over-exploitation, see Michael Benton, *ibid.*

Afterword

1   All the supposedly comprehensive report says on climate is that 'this report does not attempt to address the complexities of this issue' (see http://www.epa.gov/indicators/roe/html/roeAirGlo.htm).

2   Palutikof, J., 2003: 'Global temperature record', Climatic Research Unit Information Sheet (http://www.cru.uea.ac.uk/cru/info/warming/).

3   WMO, 2004: 'WMO statement on the status of the global climate in 2003'.

4   Hopkin, M., 2004: 'Extreme heat on the rise', *Nature*, doi: 10.1038/nature02300 (2004).

5   Krajick, K., 2004: 'All downhill from here?' *Science*, 303, 1600–1602.

6   Krajick, K., 2004: *Ibid.*

7   McCarthy, M., 2004: 'Disaster at sea: global warming hits UK birds',
    Independent (30 July 2004) and Kettlewell, J., 2004: 'Climate warning
    from the deep', BBC News Online (12 July 2004,
    http://news.bbc.co.uk/1/hi/sci/tech/3879841.stm).

8   Parmesan, C., and Yohe, G., 2003: 'A globally coherent fingerprint
    of climate change impacts across natural systems', *Nature*, 421
    (2 January 2003), 37–42, and also Root *et al.*, 2003: 'Fingerprints
    of global warming on wild animals and plants', *Nature*, 421
    (2 January 2003), 57–60.

9   BBC News Online, 2004: 'Global Warning'
    (http://newsvote.bbc.co.uk/1/hi/in_depth/sci_tech/2004/
    climate_change/default.stm).

10 AFP, 2004: 'Pacific's low-lying Tuvalu braced for more "king waves"'.

11 Reuters, 2004: 'Dust storm blankets Chinese capital' (30 March
    2004).

12 Associated Press, 2004: 'Peru warns of water rationing in capital'
    (10 March 2004).

13 Vargas, M., 2004: 'Global warming melts Peruvian peaks', Reuters
    (26 July 2004).

14 Franzen, O., 2004: personal communication.

15 See http://www.cpc.ncep.noaa.gov/products/outlooks/figure3.gif

Illustration Credits

Index

P.S.

Ideas,
interviews
& features ...

In Conversation

Mark Lynas in conversation with Mayer Hillman, author (with Tina Fawcett) of *How We Can Save the Planet* (Penguin, 2004), October 2004.

ML: It's an historic week because the Russians have decided to ratify Kyoto, which I think is a very valuable first step in the right direction – especially as everyone, even the EU, is currently well short of their national targets on lowering carbon emissions. So there will be some kind of real-world reckoning even with this limited treaty.

MH: But as you know, Kyoto only applies to the developed world. I think it's been overtaken by the Global Commons Institute's proposal for *Contraction and Convergence*, which is commanding wider and wider international support and will replace Kyoto. Kyoto is dead in the water. The Americans have been hoisted by their own petard because they demanded the participation of developing countries. Well, *C&C*, based on equal per capita emissions across the world, fits that bill perfectly.

ML: But that won't bother the Americans. They were only ever citing it as a stalling technique, weren't they?

MH: It doesn't matter what they were doing it for. When we have carbon rationing – and it's going to be soon because there isn't any other assured way of saving the planet from excess greenhouse gases – we will be able to understand our common, shared responsibilities. There can be only two implications of our continuing to exceed our ration: either we've got to stop other people – particularly in the Third World – from having their ration, or we don't really care about the future condition of the planet. What else can happen?

ML: You're several steps along there – people actually have to accept the need for carbon limitation in the first place. You walk down the street and you see people living their lives, driving cars, going to the supermarket, doing whatever people do, and it's clearly not even in the top hundred of most people's level of concerns. Given that we need serious political action within ten years or so, the problem is that we don't have time for the public to wake up.

MH: Absolutely. These changes have got to be made, and soon. The public haven't begun to appreciate the extent of the change that is in prospect. But we shouldn't criticize them for this – yet – because they're causing damage *unwittingly*. Once they understand the nature of the situation, they will be doing it *wittingly*. Their next step is then to act rationally and responsibly.

ML: But where do you make the divide then? Even my closest friends and family know what's going on because they've read my book and we've talked about it ad nauseam, but I still catch them booking holidays to Barbados . . . ▶

Mark Lynas

❝ We don't have time for the public to wake up. ❞

In Conversation *(continued)*

◀ **MH:** What's their line of defence?

ML: They just switch off, or they dismiss you and say 'Oh, there you go again.'

MH: You're right. We're dealing here with the psychology of denial. I find it quite remarkable that people either change the subject or joke about its seriousness.

ML: The joke approach I think is interesting, because I like humour and often have a slightly humorous perspective on life. But when people make jokes about global warming, I find it irritating – it's almost a way of not facing up to the issue. I have a sense of humour failure!

MH: Exactly so. You wouldn't joke about the Beslan massacre. A World Health Organization report estimated earlier in the year that 160,000 people died in the developing world due to excess heat attributable to global warming in 2003. You wouldn't have people laughing about that. There is a need for confrontation – even with close friends and family. You can't let people get away with it by joking or changing the subject. This is the harsh reality: that, as a consequence of you carrying on as you are, climate change will be made worse. What are you going to do about it? Because you are now complicit in this worsening process.

ML: Are you not too bothered about the social pressure, then? I don't like to stand up

> ❛ 160,000 people died in the developing world due to excess heat attributable to global warming in 2003. ❜

in a situation and feel like I'm embarrassing everybody and am causing a nuisance.

MH: No. Here's an analogy: SS troops making Jews wash the pavements in Vienna in 1938. As you pass by, you feel it would be embarrassing to make a scene, although you know that you should.

ML: It's that denial thing again.

MH: Absolutely. Your book has set down terrible things going on now. You've seen it at first hand – equivalent to seeing Jews being obliged to wash pavements, except far worse.

ML: What I've seen is worse than that?

MH: Oh, far worse. You have seen evidence of changes occurring in the ecology of the planet that bode extremely ill for the survival and quality of life of future populations. We're accountable to our children and grandchildren, who'll stand over our graves in 2050 and say: 'What were you doing, fifty years ago, when all the evidence was there, and you still preferred to carry on? Look at the devastation around us. Look at the way our lives have been blighted by your selfishness. How could you do it?'

ML: No one should feel able to excuse themselves on the grounds that they didn't know. But what do you have to do to really convince people? I almost feel like the number of us who realize the gravity of the situation is probably in the low hundreds ▶

‘We're accountable to our children and grandchildren.’

◀ in the whole of the country. It's almost like we're a religious group, wearing our sandwich boards and so on. And that's why people glaze over, because they realize that we're trying to preach at them about something and that our belief is so strong that we're no longer able to be rational about it in their eyes.

MH: I don't think they do think that actually. I think they're just non-plussed.

ML: Or people say: 'I think it's great what you're doing, and I really like the book,' and I reply: 'Thanks – what are you doing to change your lifestyle?' and there's no reply.

MH: That's right. I think we're agreed that this is the heart of the matter: how do we get over people not wanting to know because it may require them to stop doing what they *prefer* to do. I don't think the penny has dropped that we are about to witness the end of capitalism. The need for carbon rationing simply reflects the fact that we can no longer pursue GDP as the measure of progress because it is too closely coupled to consuming resources and accelerating the process of harmful climate change.

ML: I still rather feel that if you say things like that, you're making the task look even harder than it actually is. Because the majority of society is so bound up with capitalism – that's how people live their lives, how everything is structured socially, that's how the economy works and so on. And whilst I

❛ I don't think the penny has dropped that we are about to witness the end of capitalism. ❜

6

would love to see the end of capitalism, that's almost a harder task than facing up to climate change, isn't it?

MH: The bottom line is the protection of the global environment. Maybe it's a failure on my part, but I just can't understand how anybody cannot but accept the simplicity of the argument: here we've a planet with a finite capacity to absorb greenhouse gases. Divide these by the world population to get an average share of emissions set at a level to prevent serious damage to the ecology of the planet. The climate scientists almost unanimously say that this global average has got to be reduced by at least 60 per cent, so you get down to about 1 tonne per person. How anybody can just say or imply by their action or inaction 'I don't want to know' is beyond me.

ML: I'll tell you how they get out of it. One word: technology. They just imagine that if you run enough cars with hydrogen fuel cells and build enough wind turbines we'll solve the problem.

MH: That's right. But all the evidence points to the fact that you can't just substitute our energy-intensive lifestyles with renewables and energy efficiency – it simply can't be done.

ML: Why not?

MH: Because there isn't the capacity to do so – within the limited timetable open to us, ▶

‘ All the evidence points to the fact that you can't just substitute our energy-intensive lifestyles with renewables and energy efficiency. ’

In Conversation *(continued)*

◀ at a cost that we can afford, and at the same time continuing in the pursuit of economic growth. The demands are too high. You can see that, over the last ten or twenty years, the benefits of improved technology have generally been outpaced by the growth in demand for energy. In the main, technology has enabled some decoupling of growth from resource use. But we can't have growth – certainly not in the form in which we now measure it. We have twenty years at most – the world is already going to be devastated by virtue of what we've done up to 2004. It will be even worse if we don't act now.

> ❛ The world is already going to be devastated by virtue of what we've done up to 2004. It will be even worse if we don't act now. ❜

ML: But has it not been said that if some entrepreneur were to build solar panels over half of the Sahara desert, that would be enough power to run the world? That's a technological magic bullet. Are you ruling that out?

MH: No. The role that technology plays in combating climate change will be the role that it can play and that it is encouraged to play against the background of people living within their carbon ration. The importance of the imposition of rationing is to *assuredly* protect the planet. Simply proposing solar panels in the Sahara won't achieve that. With carbon rationing, if people find that solar panels in the Sahara are the best way for them to live within their ration, that will be the way to go.

ML: So you're saying, and I think I agree with this, that the policy framework has to come first, because that's defining the entirety of the way that society is going to have to change. But the thing is, people use technology as a reason not to set that policy framework up to start with!

MH: Absolutely. And a further problem lies in the fact that we can't rely on sufficient commitment from people who are persuaded of the need to 'behave ecologically'. My analogy is with 1939. Before you declare war, you can ask for volunteers. But if the war is adjudged in the public interest, government intervention in the form of conscription can be seen to be justified if there are insufficient volunteers.

ML: So you can really foresee a situation where the people as a whole actually ask the government to have something imposed on them, like carbon rationing, that is going to change their lifestyles? At the moment I can't see it. We go back to the problem where the public isn't even one hundredth of the way along this road. I know there are people in government who are much more advanced, but I can't see any circumstances in which they'd have the political courage to float this as a policy. It's inconceivable.

MH: Well then, go back to August 1939. Would you not have been wheeling out those arguments against the proposition that food rationing should be introduced?

ML: Okay, so how do we get war declared? Swap the war on terror for a war on global ▶

> ❛ We can't rely on sufficient commitment from people who are persuaded of the need to "behave ecologically". ❜

In Conversation *(continued)*

◄ warming? The thing about wars is that you have a sense of national emergency because you're imminently about to be attacked. The problem with climate change is that it's too long-term for that.

MH: Maybe I'm too emotional about these things, but when I read your chapter on Tuvalu, I was almost in tears. I think that the reason why people can get away with it is because people are not confronted with the realities of their actions. Give me anybody, whether it's Tony Blair or the man in the street, and ask: 'Do you really not care about the future of the planet?'

ML: What happens if they say yes?

MH: You're right. A few will. But the great majority of the population have either an emotional stake in the future because they've got children or they would like to feel that the world will carry on after they're gone.

ML: But you're presupposing that people will behave rationally, which they clearly don't by any measure of public behaviour. People believe all sorts of things. For example, a big majority thinks vast amounts of public money are spent on asylum seekers – none of that's true. That's probably the same for any policy you choose.

MH: Whenever I hear the argument that the task is fraught with difficulty, I say: 'Are you implying that we should give up because of that, or that we must therefore redouble our

> ❝ People are not confronted with the realities of their actions. ❞

efforts?' The way through is so obvious: we've got to inform the public that carrying on with 'business as usual' is leading to ecological Armageddon. There were no demonstrations in Trafalgar Square when food rationing was introduced in 1939 because people understood that this was the only *fair* way of coping with shortage.

ML: It's almost as if we have to find some way to create a sense of imminent threat.

MH: Well, that's what your book does! My view, and this is why I remain an optimist, is that when people are presented with the facts that we cannot continue with our energy-affluent lifestyles, they will act rationally even if that entails personal hurt. This is really at the heart of the matter. A friendship of mine of fifty years has been put at serious risk because I refused to fly to visit him in Canada. When the friend asked if I minded that he flew to the UK to visit us, I said 'Yes, I did.'

ML: But you're almost unique in the way that you're prepared to walk the walk. I mean, I wouldn't take a jet flight on holiday, but I had to in order to do my research.

MH: Would you vote for a government that said: 'Our first act will be to introduce carbon rationing'?

ML: Of course. Tomorrow.

MH: Well, that vindicates everything I've ▶

❛ We've got to inform the public that carrying on with "business as usual" is leading to ecological Armageddon. ❜

In Conversation (*continued*)

◀ said, which is that it's unrealistic to expect more than a few people to change unilaterally, but once the rationing framework is in place, it all makes sense. The reason you're prepared to welcome rationing is because you understand why this is the only way of saving the planet.

ML: People often assume individual self-interest is our only motivation, but humankind has evolved an awareness of group self-interest. We are social animals and we are prepared to make shared sacrifices on behalf of the group, which is what rationing appeals to. This reminds me that I actually feel a low-carbon world would be far preferable to the one we're currently living in. So why is it that we see giving up fossil fuels as some kind of big sacrifice? It would be better for everyone!

MH: Yes, absolutely right. We'd have local food supplies, better communities. People would be healthier. I think the so-called obesity crisis would be solved overnight by taking motorized transport out of the equation.

ML: It's like we've replaced human calories burned within the body with fossil carbon calories burned in engines, so everyone's storing up the unburned human calories into great big rolls of fat! There's a simple energy equation there.

MH: This is why children today are so disadvantaged by not getting around by bike

<blockquote>❛ Why is it that we see giving up fossil fuels as some kind of big sacrifice? It would be better for everyone! ❜</blockquote>

and on foot. More of them are becoming obese, not because they're eating more; it's because they're getting less exercise.

ML: So, we agree that carbon rationing is the only way to go, but what would it actually look like?

MH: It's simply a swipe card. It can easily be done when you pay your gas or electricity bills, buy petrol, train tickets or even fruit that has had to be transported a long distance. It is easy now with technology. It's not a problem. I've been arguing the case for carbon rationing for over fourteen years, and have listened very hard to the counter-arguments, and remain convinced that there are no practical problems with its introduction.

ML: Everything, the whole direction of society, once you've got that policy framework, will shift.

MH: Exactly. Rather than trying to persuade people to insulate their homes, they will be strongly encouraged by the consequences of not doing so. And the entrepreneurs will be knocking on their doors saying: 'Look, if you let me insulate your home I'll give you money, because your unused carbon ration is going to be worth so much.'

ML: I wonder, when it seems so obvious when you talk about it, why carbon rationing is still such a fringe issue? Even the environmental groups don't seem to have taken ▶

❮ I've been arguing the case for carbon rationing for over fourteen years. ❯

In Conversation *(continued)*

◀ this on board. Of course it only works if it's done globally as part of *Contraction and Convergence*.

MH: Well, it is clearly too ambitious to expect international agreement right away. The strategy is for the EU and Africa to take the lead soon on the adoption of *Contraction and Convergence*. Meetings are going on about that now. Even the USA doesn't have a fence around it to protect it from climate change. The devastation of drought in the Midwest for two or three years will soon bring Americans to their senses.

ML: But we're running out of time. This has always been the conundrum with climate change – that by the time the impacts are sufficiently damaging that even the worst deniers in America wake up, we'll have gone way too far down the road and will be heading for five or six degrees of devastation. We could have a global climate regime that doesn't include America, but then they'll be free-riding on our actions.

MH: If the USA wants to become a pariah state like South Africa during the period of apartheid . . . once there's trade sanctions, and bans in sporting tournaments, cultural events, and so on, they will soon fall into line.

ML: Meanwhile, we've got to persuade the government to introduce carbon rationing.

MH: We're on the way. The prime minister didn't say that climate change is the most serious threat facing the planet unless, presumably, he was persuaded that it is his responsibility – *primus inter pares* – to take the lead in the UK in limiting the damage.

ML: I hope you're right! ∎